THE MATHEMATICAL DESCRIPTION OF SHAPE AND FORM

E.A. LORD, Ph.D.
formerly Research Fellow, Department of Architecture
University of Edinburgh

and

C.B. WILSON, Ph.D.
Professor of Architectural Science
Department of Architecture
University of Edinburgh

ELLIS HORWOOD LIMITED
Publishers · Chichester

Halsted Press: a division of
JOHN WILEY & SONS
New York · Chichester · Brisbane · Toronto

First published in 1984 and
Reprinted in 1986 by
ELLIS HORWOOD LIMITED
Market Cross House, Cooper Street, Chichester, West Sussex, PO19 1EB, England

The publisher's colophon is reproduced from James Gillison's drawing of the ancient Market Cross, Chichester.

Distributors:

Australia, New Zealand, South-east Asia:
Jacaranda-Wiley Ltd., Jacaranda Press,
JOHN WILEY & SONS INC.,
G.P.O. Box 859, Brisbane, Queensland 4001, Australia

Canada:
JOHN WILEY & SONS CANADA LIMITED
22 Worcester Road, Rexdale, Ontario, Canada.

Europe, Africa:
JOHN WILEY & SONS LIMITED
Baffins Lane, Chichester, West Sussex, England.

North and South America and the rest of the world:
Halsted Press: a division of
JOHN WILEY & SONS
605 Third Avenue, New York, N.Y. 10158, U.S.A.

© 1986 E.A. Lord and C.B. Wilson/Ellis Horwood Limited

British Library Cataloguing in Publication Data
Lord, E.A.
The mathematical description of shape and form. −
(Ellis Horwood series in mathematics and its applications)
1. Geometry
I. Title II. Wilson, C.B.
516 QA445

Library of Congress Card No. 83-26685

ISBN 0-85312-722-0 (Ellis Horwood Limited)
ISBN 0-470-20043-X (Halsted Press)

Printed in Great Britain by Unwin Brothers of Woking.

Contents

6

Preface

A method of form description is a procedure for selecting and presenting information about the characteristic way in which an object occupies space. Methods of form description are, in practice, always chosen for a particular purpose, in a particular problem context. The fundamental issue involved in the choice of a method of form description is the identification of which information should be included as relevant to the problem, and what manner of presentation of that information best serves the process of solution of the problem.

We first encountered the problem of form description in the particular context of problems of prediction of the behaviour of environmental fields in interaction with buildings. A survey of mathematical methods of form description, covering a much wider problem area, was undertaken in order to gain some insight into the criteria that operate in the choice of a method of form description. The literature contains a widely scattered conglomeration of disparate and, at first sight, unrelated methods. In most situations, there are no well-established criteria for arriving at an appropriate mode of description; approaches to form description are usually highly intuitive and pragmatic *ad hoc* responses to practical needs.

There is a distinct lack of, and an urgent need for, a rigorously developed *science of morphology* which would deal with the underlying principles of form description. The *mathematics of form description*, as a possible field of study in its own right, has not hitherto been rigorously developed, or even adequately recognised.

Thus, in this survey, we are not presenting a compendium of unrelated mathematical techniques. Instead, we have attempted to present a unified view of the mathematics of form description, emphasising underlying mathematical principles. The large number of applications reviewed or cited are chosen as illustrations of those principles. Our aim is to encourage more flexible and creative mathematical thinking about problems of form, to enable a reader whose field of study involves some aspect of form description to gain a clearer perception of how his or her particular speciality fits into a broader mathematical background, and to provide a first step towards the future development of an adequate science of morphology.

Many problems of form have physical and dynamical aspects, as well as geometrical aspects. The material properties of building components belong to the 'form' of a building in a broad sense, and have to be taken into account along with the geometry in the determination of, for example, heat flow. The generation of the form of a living organism is brought about by a complex interplay of physical forces within the organism and between the organism and its environment. In order to restrict the scope of our subject matter to manageable proportions, we have chosen to concern ourselves in this work (except in a few instances) with the purely *geometrical* aspects of form.

In common speech, the words *form* and *shape* are not clearly distinguished. We have found it expedient in this work to increase the precision of these terms in the following way. We have chosen the word *shape* to indicate those aspects of geometrical form which have to do with the external aspect that an object presents to the world. The word *form* has been reserved to indicate that some aspect of internal structure is also under consideration. For example, we shall call the morphology of a physical field the *form* of the field, whereas the geometrical properties of the external surface of an object constitute its *shape*.

We assume that the reader posesses a certain degree of mathematical expertise. The readership we have in mind consists of engineers, scientists, computer programmers, etc. We presuppose a good grasp of, for example, functions, vectors, matrices, and differential equations. However, the esoteric notation and terminology employed by professional mathematicians (in topology and differential geometry, for example) has been avoided.

Because of the scope of the subject matter, the treatment of each topic is necessarily extremely concise; our intention was to achieve this without sacrificing intelligibility. What we have sacrificed, in the interests of brevity, is detailed explanation of formulae (which we take to be self-explanatory for a mathematically well-educated reader who is prepared to think hard and go slowly), and discussions of mathematical subtleties that would be demanded in a more extended and rigorous approach.

A word about the bibliography: clearly, a comprehensive listing of all the literature that could be construed as having something to do with 'mathematical form description' is out of the question. The bibliography is highly idiosyncratic, and is intended to be. The works that we have chosen to include seem to belong to the following three categories: those which have substantially assisted in the development of our own knowledge about the subject, those which treat the various branches of mathematics on which the principles of form description are based and which in our view are clearly written, and those dealing with applications which serve as representative examples of the use of those principles.

The authors wish to thank the Science and Engineering Research Council for financial support.

Acknowledgements

We are grateful to the following for permission to reproduce copyright material:

E. L. Brill for fig. 228
Camper and Nicholson (Yachts) Ltd. for fig. 26
Cambridge University Press for figs. 38, 85 and 101
W. H. Freeman and Co. for fig. 160
P. S. Stevens for figs. 132 and 133
J. P. Duncan, G. W. Vickers and the publishers of *Computer-Aided Design* for figs. 192 and 193 which first appeared in 'Simplified method for interactive adjustment of surfaces', J. P. Duncan and G. W. Vickers, *Computer-Aided Design, 12*, 6 (Nov. 1980).

In a few other cases it has proved impossible to contact copyright holders. We regret this and hope that no offence has been given.

Chapter 1
Introduction

Mathematics is a language. The use of this language for the purpose of form description tends to highlight certain aspects of form, while suppressing others; mathematics, in common with other languages, is able to express some things more easily and naturally than others. The study of the singularities of curves and surfaces (Chapter 5) is a prime example of the way in which the use of the mathematical language predisposes the user to investigate certain particular aspects of form (or shape). What is expressed by a mathematical description is an *idealised* form — an abstract concept. The extent to which this abstract concept can be usefully regarded as a description of the form of an object existing in the real world (or potentially existing in the real world, such as the design of a man-made object intended for production) depends on the extent to which characteristics of the object *correspond* to characteristics of the idealised form. Thus, the application of a mathematical mode of form description singles out certain characteristics of an object and ignores others. The suitability of the description in a particular problem context depends on whether the characteristics singled out are relevant to the problem which gave rise to the need for description. All description is *purpose-oriented*, and the choice of descriptive method must, in practical situations, be guided by the problem context. We have attempted to emphasise this aspect of form description in our choice of illustrative examples of practical applications.

Historically, methods of form description have arisen in two ways. In the first place, there is the pragmatic way, whereby an *ad hoc* descriptive method is devised as a response to a particular problem need. This process is well illustrated, for example, by the methods of curve and surface fitting (Chapter 10) and by the methods of feature selection involved in devising character-recognition programs (Chapter 11). The mathematical methods involved here would probably not have arisen out of the evolution of mathematical science in the absence of interaction with needs arising in the 'real world'.

The above process can be contrasted with the development of mathematics for its own sake, that takes place independently of practical problems. Euclid, for example, regarded his development of geometry as a self-sufficient exercise in pure thought; applications to practical problems were incidental and of lesser interest. Many branches of mathematics are fully justified by their elegance and intrinsic interest; the question 'what use is it?' is irrelevant in such cases. Without this kind of mathematics, the development of mathematical science would be severely handicapped. From a utilitarian viewpoint, the exploration of mathematics for its own sake can be justified by observing the large number of developments in mathematics which find an application only after their discovery — sometimes a long time after. The Fourier methods (Chapter 12) are a striking example. In view of these arguments, we have felt justified in including some mathematical methods of form description which *at present* have found no practical applications (some of the topology, for example). Our principal aim has been to reveal the possibilities of mathematics as a tool for exploring the nature of shape and form and for developing imaginative ways of thinking about shape and form, rather than simply to present a review of current applied methods.

Natural forms, in particular biological forms (and, in a manner less amenable to mathematical analysis, geological forms) are to a greater or lesser extent determined by the forces that have produced them. The role of physical forces in the generation of biological form has been explored by D'Arcy Thompson [T3] in his fascinating and classic work *On Growth and Form*. The same principle operates in certain man-made forms (e.g. large suspension structures such as tents), and a related principle is at work when a man-made form is designed with reference to the forces that act on it (suspension bridges, dams, aeroplanes, etc.). A thorough treatment of this aspect of form determination would venture into physics, and so lies outside the scope of our subject matter, which we have restricted to the purely *geometrical* aspects of shape and form. However, in Chapter 7, which deals with minimising principles, we have touched upon this aspect of form description.

There is a natural subdivision of our subject matter, around which the book has been structured. The idealised form associated with a mathematical method of description may be either *continuous*, as for example when a circle is described by an equation, or *discrete*, as for example when the idealised form is a finite or infinite set of separate points. We do not wish to apply these terms in a rigorous way; they nevertheless serve the useful purpose of broadly classifying descriptive methods and thus clarifying the structure of our subject matter.

In crystallography (Chapter 8) discrete description is applied to objects whose form possesses characteristics which make discrete description appropriate and natural; the forms described have properties of discreteness, the discreteness is not an artifact imposed by the mode of description. In other circumstances, the discreteness may belong only to the mode of description and not to the form described, as for instance when the shape of a continuous curve is indicated by locating a finite number of points on it, or when an image is presented to a machine, for recognition, as an array of 'grey values' (Chapter 11). In these contexts, problems of obtaining an approximate continuous description from a given discrete description arise (interpolation).

Continuous descriptions, expressed in terms of known continuous functions, arise only in very special cases. They are usually concerned with approximations in which the broad 'smoothed out' features of a shape or form are singled out because they have significance in a particular problem context. An example is the use of trend-surface analysis in cartography (Chapter 10). Numerical descriptions are by their nature discrete. Nevertheless, the principles of continuous description are important in discussing numerical methods, as a conceptual basis. A generic example is the numerical description of the topography of a piece of land by locating a finite number of points on its surface. The finite discrete description is conceived of as a description of a continuous surface (which is itself an idealised conception of the actual topography). The ideal of continuity is thus inherent in the topographical problem, even though the actual data are necessarily discrete.

Chapters 1-7 are concerned with continuous methods of description. Chapters 8-11 are concerned primarily with discrete methods of description. The Fourier methods (Chapter 12) represent a kind of hybrid.

Chapter 2

Geometry: an outline of its historical development

Geometry is the branch of mathematics which abstracts from our perceptual experience of the shapes of objects in space and the positional relationships of objects to each other.

The usefulness of mathematics as a tool for understanding the world results from the process of *abstraction* involved in mathematical description, which focuses on the general rather than the particular. For example, the formula for the volume of a cylinder is applicable to any cylinder. The earliest geometrical methods were concerned with the search for formulae of this kind, in response to practical problems of mensuration and surveying (the word 'geometry' comes from the Greek γηομετρια, 'earth-measuring'). Formulae were often empirically derived, with little understanding of general principles or the notion of mathematical proof. Euclid recognised the underlying logical principles inherent in this body of knowledge. His systematic development of those principles is undoubtedly one of the greatest achievements of the human mind.

A fascinating survey of the development of geometrical techniques, from the earliest beginnings to the early twentieth century, is Coolidge's *A History of Geometrical Methods* [C8].

In this chapter we present a (necessarily highly condensed) survey of the principal developments that have taken place in geometry since Euclid. Our aim is to convey the 'flavour' of each branch of geometry and to emphasise the unity of the subject matter by indicating the way in which the various branches are interrelated.

2.1 Congruence and the Euclidean Group

A concept of fundamental importance in Euclidean geometry is that of the *congruence* of two figures. Suppose we are given two figures in the Euclidean plane (or in a three-dimensional Euclidean space). And suppose that, for every point in one of the figures, we can identify a corresponding point in the other, in such a way that the distance between any two points in one figure is equal to the distance between the corresponding two points in the other. We then say that the two figures are *congruent* to each other.

The correspondence between two congruent figures is a particular example of a *mapping* between two sets of points.

Let A and B be two sets of points, and, for each point P in A, choose a point P′ in B. We then say that we have established a *mapping* from A to B, which associates, with every point P in A, a unique *image* P′ in B. If every image is the image of only one point, the mapping is called *one-to-one*, and if every point of B is an image, the mapping is called a mapping of A *onto* B. In an obvious sense, every one-to-one mapping of A onto B has an inverse, which is a one-to-one mapping of B onto A. In order to emphasise this two-way property of a one-to-one onto mapping, one speaks of a one-to-one correspondence between A and B. Under a one-to-one correspondence, every point of either A or B belongs to a unique pair of points, of which one is in A and the other is in B. Thus:

Two figures in a Euclidean plane (or a Euclidean 3-space) are congruent if there is a one-to-one correspondence between A and B that preserves distances.

A one-to-one mapping of a Euclidean plane onto *itself* that preserves distances is called an *isometry*. Two figures in the plane are congruent if, and only if, there is an isometry for which one is the image of the other.

Examples of isometries in a plane are: *rotation* about a fixed point, through a given angle, *translation* in a fixed direction, through a given distance, and *reflection* in a given line.

The combination of two isometries is itself an isometry. There is an *identity* isometry, namely the mapping which associates every point of the plane with itself. Every isometry has an inverse. Hence, all the isometries of the Euclidean plane constitute a group, called the *group of motions* of the Euclidean plane, or the *two-dimensional Euclidean group*.

Similarly, congruence in three-dimensional Euclidean space establishes the concept of the three-dimensional Euclidean group.

2.2 Cartesian Coordinates

The first major advance in the development of Euclidean geometry, since the time of Euclid, was Descartes' introduction of the *coordinate system*. A coordinate system for the Euclidean plane assigns to every point in the plane a pair of numbers (x, y). Geometrical problems concerning the properties of figures in the plane then become algebraic problems. In terms of a Cartesian coordinate system, the metrical properties of figures are dependent on the prescription

$$s^2 = (x_1 - x_2)^2 + (y_1 - y_2)^2 \tag{2.1}$$

for the distance between two points (x_1, y_1) and (x_2, y_2). For 'infinitesimally close' points (x, y) and $(x + dx, y + dy)$, we can write

$$ds^2 = dx^2 + dy^2 \tag{2.2}$$

This latter form is more convenient for studying the differential properties of curves, and the generalisation to curvilinear coordinate systems. We now have a structure on R_2, the set of ordered pairs of real numbers, which is equivalent to the structure of plane Euclidean geometry.

In terms of Cartesian coordinates, an *isometry* of the Euclidean plane can be described as follows: Let (x', y') be the coordinates of the image under an isometry, of the general point whose coordinates are (x, y). Since an isometry must map any straight line to a straight line, the isometry will be described by a linear relationship, of the form

$$x' = \alpha x + \beta y + a$$
$$y' = \gamma x + \delta y + b \tag{2.3}$$

The condition that distances shall be preserved ($ds' = ds$ for any two 'infinitesimally close' points and their images) then restricts the matrix

$$A = \begin{pmatrix} \alpha & \beta \\ \gamma & \delta \end{pmatrix} \tag{2.4}$$

to be *orthogonal*:

$$A^T A = 1 \tag{2.5}$$

(where the superscript T denotes matrix transpose).

It will frequently be found convenient to employ an abridged notation. We denote the Cartesian components of a general point of the plane by (x_1, x_2) instead of (x, y), and introduce a *summation convention*: if an index appears twice in an expression, a summation over all the values of that index is implied. Thus, in the present context, (2.2) can be written in the form

$$ds^2 = dx_\alpha \, dx_\alpha \tag{2.6}$$

and the linear expressions (2.3) are expressed simply by

$$x'_\alpha = a_{\alpha\beta} x_\beta + a_\beta \tag{2.7}$$

The orthogonality of the matrix is expressed by

$$a_{\alpha\gamma} a_{\beta\gamma} = \delta_{\alpha\beta} = \begin{cases} 1 \text{ when } \alpha = \beta \\ 0 \text{ otherwise} \end{cases} \tag{2.8}$$

(ie: $\delta_{\alpha\beta}$ denotes the set of components of the unit matrix).

2.3 Plane Curves

The intuitive picture of a curve as the locus of a moving point leads to the *parametric* description of a plane curve, which expresses the Cartesian coordinates of points of the curve as continuous functions of a parameter t, that takes all values in some interval and which increases as we move along the curve in a chosen direction:

$$
\left.\begin{array}{l}
x = \zeta\,(t) \\
y = \eta\,(t)
\end{array}\right\} \tag{2.9}
$$

For any portion of a curve that is nowhere parallel to the y-axis, the x-coordinate can be chosen as the parameter, and we then have the description

$$
y = f(x) \tag{2.10}
$$

An alternative description is by means of a function of two variables,

$$
\phi(x, y) = 0 \tag{2.11}
$$

which can be regarded, in principle, as being obtained by eliminating the parameter from the two equations (2.9).

A special form of the parametric description is obtained by choosing the parameter to be length s along the curve, measured from some arbitrarily chosen reference point on the curve. We then have a special form for the parametric description,

$$
\left.\begin{array}{l}
x = x\,(s) \\
y = y\,(s)
\end{array}\right\} \tag{2.12}
$$

If the two functions here are *differentiable*, the *unit tangent vector* to the curve at every point on it can be defined. Its components are $(\dot{x}, \dot{y})$, where the dot denotes differentiation with respect to s. Abrupt 'corners' on a curve are places where the functions are not differentiable. In this chapter, *we shall always assume that any functions that we write down are continuous functions, and that any derivative appearing in formulae actually exists.*

An important descriptor of the form of a plane curve is the *curvature* κ at each point on the curve, defined as the reciprocal of the radius of curvature ρ. This is the radius of the circle touching the curve at the point under consideration, about whose centre the curve is turning as it passes through the point. From Fig. 1 it is readily apparent, from considering the limit as the point P' is moved to P, that

$$
\kappa = \dot{\psi} \tag{2.13}
$$

where the dot denotes differentiation with respect to arc length and ψ is the angle between the x-axis and the tangent to the curve. Also,

$$
\dot{x} = \cos\psi, \quad \dot{y} = \sin\psi \tag{2.14}
$$

The *intrinsic equation* of a curve expresses the curvature as a function of distance along the curve:

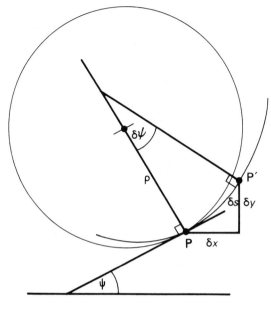

Figure 1

$$\kappa = \kappa\,(s) \tag{2.15}$$

Whereas the descriptions (2.9) and (2.11) contain information about the position and orientation of the curve relative to the coordinate axes, as well as information about the shape of the curve, the intrinsic equation describes only the shape of the curve, without reference to any coordinate system. Integrating (2.13) and (2.14) gives

$$\psi = \int \kappa\,ds, \tag{2.16}$$

$$x = \int \cos\psi\,ds, \quad y = \int \sin\psi\,ds \tag{2.17}$$

These equations provide a prescription for obtaining a parametric description of a curve (with arc length as the parameter) from an intrinsic description. The three constants of integration of course are the missing information required to specify a particular position and orientation.

2.4 Three Dimensions

If (x, y, z) are the Cartesian coordinates of a general point in a three-dimensional Euclidean space, the distance ds between two 'infinitesimally close' points is given by

$$ds^2 = dx^2 + dy^2 + dz^2 \qquad (2.18)$$

The coordinates (x, y, z) of a point are of course the components of its position vector $\mathbf{r}$, so (2.18) can be written more simply as

$$ds^2 = d\mathbf{r} \cdot d\mathbf{r} \qquad (2.19)$$

A set of points in Euclidean 3-space whose position vectors are continuous single-valued functions of two parameters,

$$\mathbf{r} = \mathbf{r}(u, v) \qquad (2.20)$$

describes a *surface*. The two families of curves u = const. and v = const. lie on the surface, and are called *coordinate curves*. Every pair of values (u, v) determines a unique point on the surface. The numbers (u, v) are the *curvilinear coordinates* of the point.

Alternative descriptions of a surface are analogous to the descriptions (2.10) and (2.11) of a closed curve. The description

$$z = f(x, y) \qquad (2.21)$$

is *Monge's form* for the description of a surface. It can be regarded as being obtained from the parametric form (2.20) by choosing x and y as parameters. A surface can also be described by a single relationship satisfied by the Cartesian coordinates of its points,

$$\phi(x, y, z) = 0 \qquad (2.22)$$

Monge's form is, of course, a special case of such a description.

The parametric description of surfaces encounters the difficulty that, for some surfaces, a single parametrisation for the whole surface does not exist. For example, if longitude and latitude are used as parameters for the surface of a sphere, the assignment of a pair of curvilinear coordinates to each point fails for the two poles. In fact, any curvilinear coordinate system on a sphere necessarily fails at at least one point. For this reason, it is necessary in general to regard a surface as covered by a set of overlapping 'coordinate patches', on each of which is imposed a different parametrisation. A different parametric description (2.20) then describes each patch of the surface, and the complete description of the whole surface consists of the collection of these separate descriptions.

The abridged notation of §2.2 is readily extended to three dimensions. Writing x_i ($i = 1, 2, 3$) for the Cartesian coordinates of a general point in a Euclidean three-space, the summation convention enables us to write (2.19) in the form

$$ds^2 = dx_i \, dx_i \tag{2.23}$$

Denoting by x_i' the coordinates of the image, under an isometry, of a point whose coordinates are x_i, the isometry is expressed by

$$x_i' = a_{ij} \, x_j + a_i \tag{2.24}$$

in which the matrix is orthogonal:

$$a_{ik} \, a_{jk} = \delta_{ij} \tag{2.25}$$

2.5 Space Curves

A parametric description of a curve in Euclidean three-space is related to the idea of a curve in 3-space as the locus of a moving point. Distance along the curve (from some arbitrarily chosen point on it) can be used as the parameter; we then have a description of the curve by an equation of the form

$$\mathbf{r} = \mathbf{r}(s) \tag{2.26}$$

Denoting differentiation with respect to s by a dot, the vector

$$\mathbf{t} = \dot{\mathbf{r}} \tag{2.27}$$

is a tangent vector to the curve at every point. By dividing (2.19) by ds^2, we see that it is a unit vector. Differentiating $\mathbf{t} \cdot \mathbf{t} = 1$ shows that $\dot{\mathbf{t}}$ is perpendicular to $\mathbf{t}$, that is, it is *normal* to the curve. The plane containing the vectors $\mathbf{t}$ and $\dot{\mathbf{t}}$ for a point on the curve is the *osculating plane* at that point. The unit vector normal to the curve, in this plane, is denoted by $\mathbf{n}$, and called the *principal normal*. We can then write

$$\dot{\mathbf{t}} = \kappa \mathbf{n} \tag{2.28}$$

The factor κ is called the *curvature* of the curve. It is clearly the curvature of the plane curve obtained by projecting the space curve onto the osculating plane at the point under consideration. The *binormal* to the curve, at any point, is a unit vector perpendicular to the osculating plane, defined by

$$\mathbf{b} = \mathbf{t} \times \mathbf{n} \tag{2.29}$$

We now have a triad of orthogonal unit vectors that changes its orientation as we move along the curve. The plane containing the vectors $\mathbf{n}$ and $\mathbf{b}$ is the *normal plane* and the plane containing $\mathbf{b}$ and $\mathbf{t}$ is the *rectifying plane*.

Since $\mathbf{n}$ is a unit vector, $\dot{\mathbf{n}}$ is perpendicular to it and is therefore a linear combination of $\mathbf{t}$ and $\mathbf{b}$. The coefficient of $\mathbf{t}$ is $\dot{\mathbf{n}} \cdot \mathbf{t} = -\mathbf{n} \cdot \dot{\mathbf{t}} = -\kappa$. Hence we can write

$$\dot{\mathbf{n}} = -\kappa \mathbf{t} + \tau \mathbf{b} \tag{2.30}$$

The coefficient τ is the *torsion* of the curve. Finally, since **b** is a unit vector, $\dot{\mathbf{b}}$ is perpendicular to it, and therefore a linear combination of **n** and **t**. The coefficient of **n** is $\dot{\mathbf{b}} \cdot \mathbf{n} = -\mathbf{b} \cdot \dot{\mathbf{n}} = -\tau$, and the coefficient of **t** is $\dot{\mathbf{b}} \cdot \mathbf{t} = -\mathbf{b} \cdot \dot{\mathbf{t}} = 0$. Therefore

$$\dot{\mathbf{b}} = -\tau\mathbf{n} \qquad (2.31)$$

The equations (2.28), (2.30) and (2.31) are the Serret-Frenet formulae [C14, D5, S20]. They describe the way in which the triad rotates as we move along the curve, and are fundamental to any study of the properties of curves in three-dimensional Euclidean space.

A knowledge of the functions $\kappa(s)$ and $\tau(s)$ determines the form of a curve uniquely, apart from its position and orientation in space. To see this, it is sufficient to note that, if the triad is given at $s = 0$ (thus fixing the orientation), the Serret-Frenet formulae can be integrated to give $\mathbf{t}(s)$. Then integration of (2.27) will give $\mathbf{r}(s)$ apart from a (vector) constant of integration which fixes the position of the curve.

Associated with every method of form description, there are certain kinds of forms whose description is particularly simple. The concept of a 'simple' shape is definable only with reference to a chosen method of description. For the intrinsic description of space curves in terms of the functions $\kappa(s)$ and $\tau(s)$, the simple shapes are curves with constant curvature and torsion. These are the *helices*. A helix is a curve which remains at a constant distance from a fixed line (the axis) and forms a constant angle with the planes perpendicular to the axis (Fig. 2). Let the axis be the z-axis, and consider the helix which passes

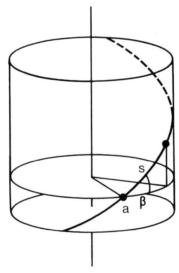

Figure 2

through the point $(a, 0, 0)$, making a constant angle β with the planes perpendicular to the axis. Then

$$\mathbf{r} = \begin{pmatrix} a \sin \varphi \\ a \cos \varphi \\ s \sin \beta \end{pmatrix} \tag{2.32}$$

where

$$\varphi = \frac{s}{a} \cos \beta \tag{2.33}$$

By repeated differentiation of $\mathbf{r}$ with respect to s, and using the Serret-Frenet formulae, it can be deduced that

$$\kappa = \frac{\cos^2 \beta}{a}, \quad \tau = \frac{\sin 2\beta}{2a}, \tag{2.34}$$

or, equivalently,

$$\frac{\tau}{\kappa} = \tan \beta, \quad \frac{\kappa}{\kappa^2 + \tau^2} = a \tag{2.35}$$

Thus every helix is a curve of constant curvature and torsion. Conversely, since curvature and torsion determine the form of a curve uniquely, every curve with constant curvature and torsion is a helix (if we include the circles $\tau = 0$ and straight lines $\kappa = 0$ as special cases).

2.6 Gaussian Surface Theory

We have already seen how a curve in the Euclidean plane or in Euclidean 3-space can be described without reference to any particular coordinate system on the space in which it is embedded. The theory underlying the analogous description of surfaces in Euclidean 3-space was worked out by Gauss [D5, C14, S20].

We start out from the parametric description (2.20), where the parameters are associated with two families of curves marked on the surface. The vector $d\mathbf{r}$ connecting two 'infinitesimally close' points on the surface is given by

$$d\mathbf{r} = \frac{\partial \mathbf{r}}{\partial u} du + \frac{\partial \mathbf{r}}{\partial v} dv \tag{2.36}$$

The distance between these points is then (from (2.19)) given by

$$ds^2 = E du^2 + 2F dudv + G dv^2, \tag{2.37}$$

where

$$E = \left(\frac{\partial \mathbf{r}}{\partial u}\right)^2, \quad F = \frac{\partial \mathbf{r}}{\partial u} \cdot \frac{\partial \mathbf{r}}{\partial v}, \quad G = \left(\frac{\partial \mathbf{r}}{\partial v}\right)^2 \tag{2.38}$$

The coefficients E, F and G, which are functions of the curvilinear coordinates (u, v) of points on the surface, are components of the *metric* or *first fundamental form* of the surface. They are independent of the particular choice of Cartesian coordinate system in which the surface may lie. The metric properties of any figure marked on the surface can be discussed in terms of the parameters (u, v) of its points, and the metric components. Thus, the metric determines the *intrinsic* geometrical properties of the surface, which are independent of whether we regard the surface as immersed in the Euclidean 3-space, or as a self-sufficient two-dimensional space existing independently of a third dimension. For example, a sheet of paper may be curled to form a portion of a cylinder or a cone, but the intrinsic geometrical properties of figures marked on it remain unchanged – the intrinsic geometry remains the geometry of a Euclidean plane. In order to specify the way a surface is immersed in Euclidean 3-space, and thus to complete the description of its shape as an object in that space, we require some additional information to supplement the intrinsic description provided by the metric. This information comes from a consideration of the way the *normal* to the surface varies from point to point on the surface.

The vector

$$\frac{\partial \mathbf{r}}{\partial u} \times \frac{\partial \mathbf{r}}{\partial v} \tag{2.39}$$

is clearly normal to the surface at each point (since $\partial \mathbf{r}/\partial u$ and $\partial \mathbf{r}/\partial v$ are tangential). The modulus of the vector (2.39) is given by

$$\left| \frac{\partial \mathbf{r}}{\partial u} \times \frac{\partial \mathbf{r}}{\partial v} \right|^2 = \left(\frac{\partial \mathbf{r}}{\partial u} \right)^2 \left(\frac{\partial \mathbf{r}}{\partial v} \right)^2 - \left(\frac{\partial \mathbf{r}}{\partial u} \cdot \frac{\partial \mathbf{r}}{\partial v} \right)^2 = EG - F^2 \tag{2.40}$$

so the unit normal $\mathbf{N}$ is given by

$$\mathbf{N} = \frac{\dfrac{\partial \mathbf{r}}{\partial u} \times \dfrac{\partial \mathbf{r}}{\partial v}}{\sqrt{(EG - F^2)}} \tag{2.41}$$

The difference in the unit normals at two infinitesimally close points on the surface is

$$d\mathbf{N} = \frac{\partial \mathbf{N}}{\partial u} du + \frac{\partial \mathbf{N}}{\partial v} dv \tag{2.42}$$

The *second fundamental form* for the surface is defined by

$$\phi = d\mathbf{N} \cdot d\mathbf{r} = D \, du^2 + 2D' \, dudv + D'' \, dv^2 \tag{2.43}$$

where

$$D = \frac{\partial N}{\partial u} \cdot \frac{\partial r}{\partial u} = -N \cdot \frac{\partial^2 r}{\partial u^2}$$

$$D' = \frac{\partial N}{\partial u} \cdot \frac{\partial r}{\partial v} = \frac{\partial N}{\partial v} \cdot \frac{\partial r}{\partial u} = -N \cdot \frac{\partial^2 r}{\partial u \partial v}$$ (2.44)

$$D'' = \frac{\partial N}{\partial v} \cdot \frac{\partial r}{\partial v} = -N \cdot \frac{\partial^2 r}{\partial v^2}$$

(The equality of the different expressions for these components of the second fundamental form comes from differentiating the identities $N \cdot \partial r/\partial u = 0$ and $N \cdot \partial r/\partial v = 0$ with respect to u and v.)

Like the components of the first fundamental form, the components D, D′, D″ of the second fundamental form are independent of the choice of the Cartesian coordinate system in the 3-space in which the surface lies. It can be shown [S20], though we shall not do so here, that a knowledge of the first and second fundamental forms determines the shape of the surface uniquely (but not, of course, its position and orientation).

In order to proceed further, we introduce some standard abbreviated notation. The parameters will be denoted by u^1 and u^2 instead of u and v. The curvilinear coordinates of a point on the surface are then u^α $(\alpha = 1, 2)$. Differentiation with respect to these parameters will be denoted by a subscript — for example, $r_\alpha = \partial r/\partial u^\alpha$. We shall use the *summation convention*: if an index occurs twice in an expression, once as a subscript and once as a superscript, a summation over that index is implied. For example,

$$dr = r_\alpha du^\alpha = \frac{\partial r}{\partial u} du + \frac{\partial r}{\partial v} dv$$ (2.45)

The components of the two fundamental forms will be denoted by

$$g_{11} = E, \quad g_{12} = F, \quad g_{22} = G,$$
$$\kappa_{11} = D, \quad \kappa_{12} = D', \quad \kappa_{22} = D''$$ (2.46)

so that the expressions (2.37) and (2.43) for the two fundamental forms are now

$$ds^2 = g_{\alpha\beta} du^\alpha du^\beta,$$
$$\phi = \kappa_{\alpha\beta} du^\alpha du^\beta$$ (2.47)

where

$$g_{\alpha\beta} = r_\alpha \cdot r_\beta, \quad \kappa_{\alpha\beta} = N_\alpha \cdot r_\beta = -N \cdot r_{\alpha\beta}$$ (2.48)

The determinant $EG - F^2$ of the matrix of components of the metric will be denoted by g, and the components of the inverse of the matrix will be denoted by $g^{\alpha\beta}$ (that the matrix is positive definite and therefore non-singular follows from the form of (2.47)). Then

$$g_{\alpha\beta}g^{\beta\gamma} = \delta^\gamma_\alpha = \begin{cases} 1 \text{ if } \alpha = \gamma \\ 0 \text{ otherwise} \end{cases} \tag{2.49}$$

The matrices $g_{\alpha\beta}$ and $g^{\alpha\beta}$ will be used to raise and lower indices. For example, the two vectors $\mathbf{r}^\alpha$ are defined by

$$\mathbf{r}^\alpha = g^{\alpha\beta}\mathbf{r}_\beta, \quad \mathbf{r}_\alpha = g_{\alpha\beta}\mathbf{r}^\beta \tag{2.50}$$

Any vector at a point on the surface can be projected onto the three reference vectors $\mathbf{r}_\alpha$, $\mathbf{N}$:

$$\mathbf{V} = V^\alpha\mathbf{r}_\alpha + V_0\mathbf{N}, \tag{2.51}$$

where

$$V^\alpha = \mathbf{r}^\alpha \cdot \mathbf{V}, \quad V_0 = \mathbf{V} \cdot \mathbf{N} \tag{2.52}$$

Applying this concept to the second derivatives of the position vector and the first derivatives of the unit normal, we find

$$\left.\begin{aligned} \mathbf{r}_{\alpha\beta} &= \begin{Bmatrix} \gamma \\ \alpha\beta \end{Bmatrix}\mathbf{r}_\gamma - \kappa_{\alpha\beta}\mathbf{N}, \\ \mathbf{N}_\alpha &= \kappa^\gamma_\alpha\mathbf{r}_\gamma \end{aligned}\right\} \tag{2.53}$$

where

$$\begin{Bmatrix} \gamma \\ \alpha\beta \end{Bmatrix} = \mathbf{r}^\gamma \cdot \mathbf{r}_{\alpha\beta} = \tfrac{1}{2}g^{\gamma\rho}(g_{\rho\alpha \cdot \beta} + g_{\beta\rho \cdot \alpha} - g_{\alpha\beta \cdot \rho}) \tag{2.54}$$

The equations (2.53) are known as the Gauss-Weingarten relations. The coefficients (2.54), formed from the metric components and their derivatives, are the *Christoffel symbols*.

Differentiation of the Gauss-Weingarten relations leads to

$$\mathbf{r}_{\alpha\beta\gamma} = \left(\begin{Bmatrix} \rho \\ \alpha\beta \end{Bmatrix}_\gamma + \begin{Bmatrix} \sigma \\ \alpha\beta \end{Bmatrix}\begin{Bmatrix} \rho \\ \sigma\gamma \end{Bmatrix} - \kappa_{\alpha\beta}\kappa^\rho_\gamma\right)\mathbf{r}_\rho - \left(\kappa_{\alpha\beta \cdot \gamma} + \begin{Bmatrix} \sigma \\ \alpha\beta \end{Bmatrix}\kappa_{\sigma\gamma}\right)\mathbf{N}, \tag{2.55}$$

$$\mathbf{N}_{\alpha\beta} = \left(\kappa^\rho_{\alpha \cdot \beta} + \kappa^\sigma_\alpha\begin{Bmatrix} \rho \\ \sigma\beta \end{Bmatrix}\right)\mathbf{r}_\rho \tag{2.56}$$

Since $\partial^2/\partial u\partial v = \partial^2/\partial v\partial u$, we have $\mathbf{r}_{\alpha\beta\gamma} = \mathbf{r}_{\alpha\gamma\beta}$ and $\mathbf{N}_{\alpha\beta} = \mathbf{N}_{\beta\alpha}$. Equating the coefficients of $\mathbf{r}_\rho$ and $\mathbf{N}$ in these expressions gives

$$R_{\beta\gamma\alpha}{}^\rho = \kappa_{\alpha\beta}\kappa^\rho_\gamma - \kappa_{\alpha\gamma}\kappa^\rho_\beta, \tag{2.57}$$

where

$$R_{\beta\gamma\alpha}{}^{\rho} = \left\{ {\rho \atop \alpha\beta} \right\}_{\gamma} - \left\{ {\rho \atop \alpha\gamma} \right\}_{\beta} + \left\{ {\sigma \atop \alpha\beta} \right\} \left\{ {\rho \atop \sigma\gamma} \right\} - \left\{ {\sigma \atop \alpha\gamma} \right\} \left\{ {\rho \atop \sigma\beta} \right\}, \tag{2.58}$$

and

$$\kappa_{\alpha\beta\cdot\gamma} - \kappa_{\alpha\gamma\cdot\beta} + \left\{ {\sigma \atop \alpha\beta} \right\} \kappa_{\sigma\gamma} - \left\{ {\sigma \atop \alpha\gamma} \right\} \kappa_{\sigma\beta} = 0 \tag{2.59}$$

The equations (2.57) are the *equations of Gauss* and the equations (2.59) are *the equations of Codazzi*. The quantities (2.58) are the components of the *curvature tensor* of the surface. The Gauss–Codazzi equations are necessary and sufficient conditions for six given functions E, F, G, D, D', D'' of u and v to be the components of the first and second fundamental forms of some surface.

We have already mentioned that two surfaces may have the same intrinsic geometrical properties and yet have different shapes as objects in three-dimensional space. We gave the example of a portion of a plane, a cylinder, and a cone. We now give another, less trivial, example.

Let ρ, φ and z be the usual cylindrical coordinates in Euclidean space. The parametric equation

$$\mathbf{r} = \begin{pmatrix} \rho \cos \varphi \\ \rho \sin \varphi \\ \varphi \end{pmatrix} \tag{2.60}$$

describes a *helicoid*. It is the shape of the ceiling of a spiral staircase [C14]. The parametric curves associated with the parameters $u = \rho$ and $v = \varphi$ are helices winding around the z-axis and straight lines perpendicular to the z-axis. The metric coefficients are easily found from (2.38). They are

$$\begin{pmatrix} E & F \\ F & G \end{pmatrix} = \begin{pmatrix} 1 & 0 \\ 0 & 1 + u^2 \end{pmatrix} \tag{2.61}$$

The parametric equation

$$\mathbf{r} = \begin{pmatrix} \rho \cos \varphi \\ \rho \sin \varphi \\ \cosh^{-1} \rho \end{pmatrix} \tag{2.62}$$

is the surface of revolution of a catenary. We shall call this surface a *catenoid*. We take as parameters $u = \sqrt{\rho^2 - 1} = \sinh z$, $v = \varphi$. The parametric curves are circles and catenaries. The prescription (2.38) for the metric components gives the same expressions (2.61) that we found for the helicoid. Thus, these two surfaces are identical from the point of view of intrinsic metric properties of figures drawn on them. Their different shapes are manifestations of their different second fundamental forms. The unit normal is

$$\mathbf{N} = \mathbf{r}_1 \times \mathbf{r}_2 / |\mathbf{r}_1 \times \mathbf{r}_2| \qquad (2.63)$$

and the prescription (2.44) leads to

$$\begin{pmatrix} D & D' \\ D' & D'' \end{pmatrix} = \begin{pmatrix} 0 & (1 + u^2)^{-\frac{1}{2}} \\ (1 + u^2)^{-\frac{1}{2}} & 0 \end{pmatrix} \qquad (2.64)$$

for the helicoid, and

$$\begin{pmatrix} D & D' \\ D' & D'' \end{pmatrix} = \begin{pmatrix} (1 + u^2)^{-1} & 0 \\ 0 & -1 \end{pmatrix} \qquad (2.65)$$

for the catenoid.

The two surfaces, and their parametric curves, are indicated in Fig. 3.

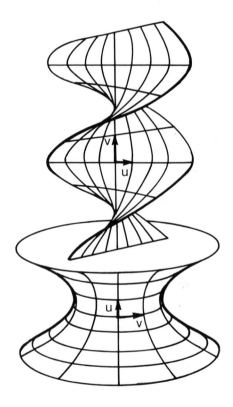

Figure 3

The description of surfaces in terms of their first and second fundamental forms finds practical application in structural engineering, when the need arises of calculating the forces acting in a thin shell or membrane structure [F2, K7].

2.7 Gaussian and Mean Curvatures

A plane passing through a point P on a surface and containing the normal to the surface at P intersects the surface in a plane curve. Denote the curvature at P, of this curve, by κ. If we consider the set of all planes through P, containing the normal at P, we obtain a set of values for κ. The maximum and minimum values of κ are the *principal curvatures* of the surface at P.

The tangent vector at P, to the curve in a normal plane, is a unit vector which can be expressed as a linear combination of the tangent vectors $\mathbf{r}_\alpha$ to the surface:

$$\mathbf{t} = \gamma^\alpha \mathbf{r}_\alpha, \tag{2.66}$$

$$\gamma^\alpha \gamma_\alpha = 1 \tag{2.67}$$

From (2.28) and (2.48), we obtain, for the curvature,

$$\kappa = \mathbf{N} \cdot \mathbf{t} = \gamma^\alpha \gamma^\beta \kappa_{\alpha\beta} \tag{2.68}$$

To find the principal curvatures, we have to find the maximum and minimum values of κ as the plane rotates about the normal; that is, as γ^α is varied subject to the restriction (2.67). We obtain

$$(\kappa_{\alpha\beta} - \lambda g_{\alpha\beta}) \gamma^\beta = 0 \tag{2.69}$$

or

$$\kappa^\alpha_\beta \gamma^\beta = \lambda \gamma^\beta \tag{2.70}$$

Thus, the *principal directions* $\gamma^\beta \mathbf{r}_\beta$ that correspond to a maximum or minimum of the curvature κ are obtained by taking the γ^β to be eigenvectors of the matrix κ^α_β. Multiplying (2.70) by γ_β and comparing with (2.68) shows that the corresponding eigenvalues are the required principal curvatures κ_1 and κ_2. If κ_1 and κ_2 are different, the two principal directions are orthogonal to each other:

$$(\kappa_1 - \kappa_2)(\gamma^\alpha_1 \mathbf{r}_\alpha) \cdot (\gamma^\beta_2 \mathbf{r}_\beta) = (\kappa_1 - \kappa_2) \gamma^\alpha_1 \gamma^\beta_2 g_{\alpha\beta} = (\kappa_1 \gamma^\alpha_1) \gamma_{2\alpha} - (\kappa_2 \gamma^\alpha_2) \gamma_{1\alpha}$$

$$= \kappa^\alpha_\beta \gamma^\beta_1 \gamma_{2\alpha} - \kappa^\alpha_\beta \gamma^\beta_2 \gamma_{1\alpha} = \kappa_{\alpha\beta}(\gamma^\beta_1 \gamma^\alpha_2 - \gamma^\alpha_1 \gamma^\beta_2) = 0$$

(since $\kappa_{\alpha\beta} = \kappa_{\beta\alpha}$). The curves on the surface whose directions are principal directions are the *lines of curvature*. There are two families of lines of curvature, which cut each other orthogonally. If the lines of curvature are taken as co-ordinate curves for a parametrisation of the surface, the matrices of components of the two fundamental forms become diagonal matrices ($F = 0$, $D' = 0$). At a point of a surface where the two principal curvatures become equal, the principal directions are undefined. Such a point is called an *umbilic*.

The two principal curvatures at a point on the surface may have equal signs or opposite signs. The surface is called *synclastic* at a point if they have the same sign and *anticlastic* if they have opposite signs. A surface has a saddle-shaped appearance in the neighbourhood of an anticlastic point. For example, the synclastic and anticlastic regions of the surface of a torus are immediately apparent.

The *Gaussian curvature* K and *mean curvature* M of a surface are the product and the mean, respectively, of the principle curvatures at a point. Synclastic and anticlastic regions correspond to positive or negative Guassian curvature. Since the principal curvatures are the eigenvalues of the matrix κ_β^α, the Gaussian and mean curvatures are the determinant, and half the trace, respectively, of that matrix:

$$K = \kappa_1 \kappa_2 = |\kappa_\beta^\alpha| = \frac{DD'' - (D')^2}{EG - F^2} \tag{2.71}$$

$$M = \tfrac{1}{2}(\kappa_1 + \kappa_2) = \kappa_\alpha^\alpha = \frac{GD - 2FD' + ED''}{EG - F^2} \tag{2.72}$$

The concept of Gaussian curvature allows the Gauss equation (2.57) to be considerably simplified. First note that we can define, from the curvature tensor (2.58), a *curvature scalar*

$$R = R_{\alpha\beta\gamma}{}^\alpha g^{\beta\gamma} \tag{2.73}$$

Since $R_{\alpha\beta\gamma\delta} = -R_{\beta\alpha\gamma\delta} = R_{\gamma\delta\alpha\beta}$, the only non-vanishing components of the curvature tensor are R_{1212} and those obtained from it by these permutations of indices. It follows that the *single* equation

$$R = \kappa_\rho^\sigma \kappa_\sigma^\rho - \kappa_\sigma^\sigma \kappa_\rho^\rho \tag{2.74}$$

contains all the information that was contained in the Gauss equation (2.57). As is easily verified, the right-hand side of (2.74) is just -2 times the determinant of κ_β^α. Therefore

$$-\tfrac{1}{2} R = K \tag{2.75}$$

This equality (apart from the numerical factor) of the curvature scalar and the Gaussian curvature is remarkable, since the curvature scalar is formed only from the metric components and their derivatives, and so is an *intrinsic* property of the surface. Therefore, the Gaussian curvature is an intrinsic property of a surface — a fact which was not at all apparent from its definition. We can infer, for example, that the Gaussian curvatures of the helicoid and the catenoid discussed in the previous section are equal. Using (2.71) to work them out, we find they are both equal to $-(1 + u^2)^{-2}$ (in fact, the mean curvatures are also equal — they are both zero — but this should be regarded as a coincidence).

The Gaussian curvature can be expressed as a scalar triple product,

$$K = \frac{1}{\sqrt{g}} [N \ N_1 \ N_2]$$ (2.76)

To prove this, note that, according to the second equation (2.53),

$$N_1 \times N_2 = \kappa_1^\alpha \kappa_2^\beta r_\alpha \times r_\beta$$

Now, equation (2.63), together with the trivial identities $r_1 \times r_1 = r_2 \times r_2 = 0$, can be written as

$$r_\alpha \times r_\beta = \sqrt{g} \ \epsilon_{\alpha\beta} \ N,$$ (2.77)

$\epsilon_{\alpha\beta}$ being the components of the matrix $\begin{pmatrix} 0 & 1 \\ -1 & 0 \end{pmatrix}$. Therefore,

$$[N \ r_\alpha \ r_\beta] = \sqrt{g} \ \epsilon_{\alpha\beta}$$ (2.78)

and so

$$[N \ N_1 \ N_2] = \kappa_1^\alpha \kappa_2^\beta \sqrt{g} \ \epsilon_{\alpha\beta} = \sqrt{g} \ (\kappa_2^1 \kappa_2^2 - (\kappa_1^2)^2) = \sqrt{g} \ K$$ (2.79)

An important property of Gaussian curvature can be obtained with the aid of (2.76). Let dS be the area of an element of a surface in Euclidean space. All the normals to dS specify a cone of directions (obtained by drawing lines parallel to the normals through a fixed point). Let the solid angle of this cone be $d\omega$, and define it to be negative if the cone is traversed anticlockwise as the perimeter of the area element is traversed clockwise. This will occur if the surface is anticlastic at the element (Fig. 4).

The Gaussian curvature on the element is then

$$K = \frac{d\omega}{dS}$$ (2.80)

Proof. Consider a small parallelogram-shaped area element bounded by co-ordinate curves. The area is

$$dS = \sqrt{g} \ dudv$$ (2.81)

The four normals at the corners can be written as $N, N + N_1 \ du, N + N_2 \ dv$ and $N + N_1 \ du + N_2 \ dv$. The solid angle $d\omega$ is therefore equal to the area of the parallelogram in Fig. 5, namely

$$d\omega = \pm |N_1 \times N_2| \ dudv,$$ (2.82)

or, since N is perpendicular to N_1 and N_2,

$$d\omega = \pm [N \ N_1 \ N_2] \ dudv = KdS$$ (2.83)

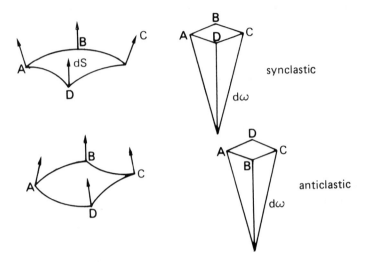

synclastic

anticlastic

Figure 4

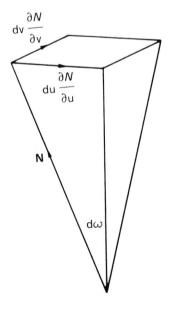

Figure 5

2.8 Curvilinear Coordinates, and non-Euclidean spaces

In many applications, the Cartesian coordinate systems for the Euclidean plane or Euclidean three-space are inconvenient. Jacob Bernoulli introduced the concept of plane polar coordinates in 1691. Other special coordinate systems were then evolved, adapted to the needs of specific physical problems, for example elliptic coordinates and (in three dimensions) cylindrical coordinates and spherical polar coordinates. Morse and Feshbach [M13] give a discussion of the various special coordinate systems that have been used in three-dimensional situations. The coordinate system is usually chosen to fit the form of a body with which a physical field interacts.

A Euclidean plane is of course a special kind of surface; it is a surface whose Gaussian curvature is everywhere zero. Thus, in terms of a pair u^α ($\alpha = 1$, 2) of curvilinear coordinates for a Euclidean plane, distance between infinitesimally close points is expressed by

$$ds^2 = g_{\alpha\beta} du^\alpha du^\beta \tag{2.84}$$

(For example, $ds^2 = dr^2 + r^2 d\theta^2$ for polar coordinates).

The fact that we are dealing with a plane, rather than a general surface, expresses itself in a set of restrictions on the metric components, namely the equation of Gauss (2.75), which now takes the form

$$R = 0 \tag{2.85}$$

Similarly, in a three-dimensional Euclidean space with a curvilinear coordinate system u^i ($i = 1, 2, 3$), we have

$$ds^2 = g_{ij} du^i du^j \tag{2.86}$$

The metric components g_{ij} satisfy a set of restrictions analogous to (2.85). In fact, according to (2.19), g_{ij} must have the special form

$$g_{ij} = \frac{\partial \mathbf{r}}{\partial u^i} \cdot \frac{\partial \mathbf{r}}{\partial u^j} \tag{2.87}$$

A necessary and sufficient condition for this is the set of six equations

$$R_{ij} = R^k_{ijk} = 0 \tag{2.88}$$

where R^l_{ijk} are defined by the three-dimensional analogue of (2.58).

The idea of studying geometries more general than Euclidean geometries, based on the fundamental relation (2.86) but with (2.88) abandoned, was put forward by Riemann in 1854, in a lecture entitled "Über die Hypothesen welche der Geometrie zu Grunde liegen" (to qualify for the post of lecturer in Gauss's department in the University of Göttingen). The quantities R^l_{ijk} are called the components of the Riemann tensor. A Riemannian geometry of any number of dimensions is a generalisation of the intrinsic geometry of a surface, which is the two-dimensional case. Euclidean space has the property of being homogeneous and isotropic, that is, its geometrical properties do not vary from point to point and are independent of orientation. There are just three kinds of space of this kind. They are the spaces of constant curvature. The coordinate system in a space of constant curvature can be chosen so that

$$ds^2 = \frac{du^\alpha du^\alpha}{1 + Ku^\beta u^\beta/4} \tag{2.89}$$

If the constant K (the curvature of the space) is zero the space is Euclidean. The two-dimensional space of constant positive curvature is the surface of a sphere; the special coordinate system which gives the metric the form (2.89) is the stereographic system obtained by projecting Cartesian coordinates of a plane tangential to the 'south pole' onto the sphere, the point of projection being the 'north pole'. A two- or three-dimensional space of constant negative curvature is the space of the 'non-Euclidean geometry' discovered independently by Gauss, Bolyai and Lobachevski in the first half of the nineteenth century (see Coxeter's *Non-Euclidean Geometry* [C13]).

The prestige of Euclid's work, and its domination of the minds of subsequent geometers, was so great that the enormous period of two thousand years was to elapse before the realisation dawned that Euclid's geometry was not the only possible logically consistent metric geometry. This is particularly remarkable in view of the fact that 'spherical trigonometry', the study of great circles on the earth's surface, is itself a non-Euclidean geometry.

Since the subject matter of this work is the description of the shapes of objects in the real world, we shall not subsequently have to deal with non-Euclidean geometries of more than two dimensions.

2.9 Projective Geometry

We now go back in history and outline a different development of geometrical ideas, which seems at first sight to have little connection with the developments we have been describing.

The artists and architects of the renaissance were faced with the following problem of form description. Given an object in three-dimensional Euclidean space, such as a building, how can we produce, by geometrical construction, a two-dimensional drawing on a plane which will represent the visual impression of the object when viewed from the proper viewing point? Clearly, the points of 3-space have to be mapped onto the picture plane by connecting them by rays to a fixed point (the *point of projection*) and determining where these rays intersect the picture plane [V4, W3]. Such a projection is called a *perspectivity*. A special case of this is *parallel projection*, for which the point of projection is infinitely distant and the rays are parallel. This special case applies to the drawing of elevations, and to *sciagraphy*, which deals with the determination of the positions of shadows cast by the sun (which is virtually 'at infinity').

The problem, arising from the science of perspective, that interested geometers was the investigation of the properties of geometrical figures in a plane which remain unchanged when the plane is mapped onto another by a perspectivity (Fig. 6). This field of investigation is plane projective geometry [C12, S6]. Clearly, this kind of geometry is not a 'metric' geometry – the concept of distance between points has no meaning in projective geometry. Nor do the concepts of angle, and of parallel lines. All quadrilaterals are equivalent in the projective plane, and all conics are equivalent (hyperbolae, parabolae and ellipses being interconvertible by perspectivities). Projective geometry is a *generalisation* of Euclidean geometry, rather than an alternative to it like Lobachevskian geometry. Any theorem that is true in projective geometry is also true in Euclidean geometry. An intermediate generalisation is *affine geometry*, which deals with geometrical properties that remain unchanged under *parallel* projection from one plane to another. For an affine plane, all parallelograms are equivalent.

A method of coordinates for plane projective geometry, analogous to the Cartesian coordinates for the Euclidean plane, consists of assigning *three* 'homogeneous' coordinates (x, y, z) to every point of the plane; they are required to be not all zero, and (x, y, z) and $(\rho x, \rho y, \rho z)$ correspond to the same point, ρ being any non-zero factor. A curve is described by a *homogeneous* equation

$$\phi (x, y, z) = 0 \qquad\qquad (2.90)$$

A straight line is determined, by a homogeneous linear equation.

The concept of homogeneous coordinates for a plane can be clarified as follows. Consider a Euclidean 3-space with a Cartesian coordinate system.

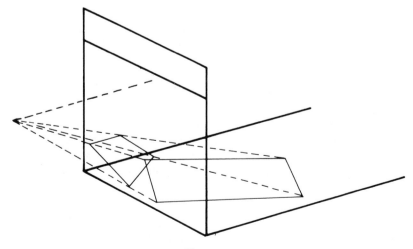

Figure 6

All the points on the line that contains the origin and a point (x, y, z) are then of the form (x, y, z). A homogeneous coordinate system on any plane not through the origin is then established by taking the Cartesian coordinates of any point on a line through the origin to be the *homogeneous* coordinates of the point of intersection of the plane with the line. It is clear from this construction that a projective plane has some extra points which the corresponding Euclidean plane does not possess – namely, those corresponding to the lines through the origin of the 3-space, *parallel* to the given plane. A projective plane can thus be regarded as obtained from a Euclidean plane by augmenting it by the inclusion of points 'at infinity', which all lie on 'the line at infinity'. The projective geometry of the plane then regards the line at infinity as in no way different from any other line. The picture we have described shows the existence of a one-to-one correspondence between all the points of a projective plane, and all the lines through a fixed point of a Euclidean 3-space.

The method of homogeneous coordinates admits of immediate generalisation to projective spaces of more than two dimensions. For example, the points of projective 3-space have four homogeneous coordinates and a surface is described by a homogeneous equation in these four coordinates.

2.10 Klein's Erlangen Program

We have indicated that Euclidean geometry is characterised by a group of isometries, which are mappings of the space onto itself that leave the geometrical properties of figures intact. We have also indicated the possibility of other geometries, such as the geometry of the surface of a sphere, Lobachevskian geometry, projective geometry.

In his inaugural address at the University of Erlangen (1872), Felix Klein [K5] introduced a unifying principle of profound importance for the understanding of what is meant by 'geometry' in its widest sense. Each kind of geometry is associated with a *group of motions**, or one-to-one mappings of the space onto itself, that leave intact the geometrical properties of figures in the space. And conversely, the specification of a group of one-to-one mappings of a space onto itself determines a geometry.

We illustrate this concept in Fig. 7, which shows a particular figure in a Euclidean plane, and its typical image under seven kinds of mapping.

(a) In Euclidean geometry, a figure and its image under an isometry are considered to be geometrically equivalent. They are called congruent.

(b) If the group of isometries is enlarged to include changes of scale (dilatations), we have the group of *similarities* of the Euclidean plane. If a figure and its

---

*Klein was actually concerned with the groups that leave the *theorems* of the geometry intact, which is sometimes a larger group than the group of motions. For example, the Klein group for Euclidean geometry is the group of similarities, not the group of isometries, and the Klein group for plane projective geometry includes 'correlations' which map points to lines and lines to points.

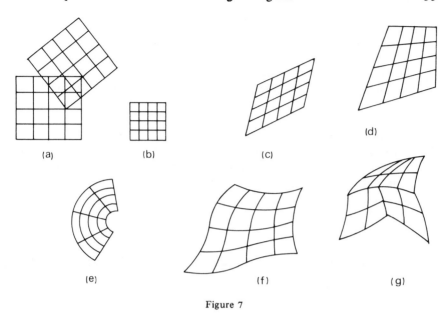

Figure 7

image are considered equivalent, we have a geometry which recognises the Euclidean concepts of angle and of length ratio, but in which the concept of length is absent.

(c) Further extension of the group by including 'shearing' transformations leads to *affine geometry*. The concepts of *parallel* lines, and length ratios of parallel line segments, are retained, but the Euclidean concepts of length and angle are absent. In affine geometry, all parallelograms are equivalent.

(d) When the group of motions consists of all transformations obtained by perspectivities, we have plane *projective* geometry. In order that such mappings shall be one-to-one mappings of the plane onto itself, the plane has to be augmented by the addition of a 'line at infinity' (which is mapped to the dotted line in the illustration). In projective geometry all quadrilaterals are equivalent.

(e) Mappings on the plane consisting of inversions in circles [C14, M11], and reflections in lines, constitute a group of one-to-one mappings of the plane onto itself if the plane is augmented by a 'point at infinity'. This is the group of motions of *inversive* geometry. The Euclidean concepts of angle and circle are retained. The concept of length, and the distinction between a line and a circle, are absent.

(f) The *differential* geometry of the plane is obtained if the group of all one-to-one differentiable mappings of the plane onto itself ('diffeomorphism') is taken to be the group of motions.

(g) If the group of motions is taken to be all *continuous* one-to-one mappings of the plane onto itself (*topological transformations*), the associated geometry is the *topology* of the plane. In this geometry, a square and a circle, for example, are equivalent, as are all continuous closed curves without self-intersections. A *topology* is a geometry which recognises only the *continuity* properties of figures.

2.11 Transformations

We conclude this chapter by indicating how the various mappings mentioned in the previous section can be described in terms of coordinates. Consider a Cartesian coordinate system on the Euclidean plane and let the coordinates of a general point be (x, y) or x^α ($\alpha = 1, 2$). The coordinates of the image under a *topological transformation* are given by a pair of *continuous* functions:

$$x'^\alpha = f^\alpha(x, y) \qquad (2.91)$$

The equations (2.91) are, of course, required to be invertible. That is, there is a pair of continuous functions g^α such that

$$x^\alpha = g^\alpha(x', y') \qquad (2.92)$$

For a *diffeomorphism*, the functions are required to be differentiable as well as continuous. The invertibility of (2.91) in this case is neatly expressed by the requirements that the Jacobian (i.e. the determinant of the matrix $\partial f^\alpha/\partial x^\beta$) should be nowhere zero.

For a *linear* transformation, (2.91) becomes

$$x'^\alpha = A^\alpha_\beta x^\beta + a^\alpha \qquad (2.93)$$

where A^α_β is a constant non-singular matrix. These are the *affine* transformations. If the matrix A^α_β is restricted to be orthogonal, we have the *Euclidean* transformations, and if it restricted to be simply a multiple of an orthogonal matrix, we have the *similarity* transformations.

The group of motions of inversive geometry is obtained by combining similarities with inversion in the unit circle,

$$x'^\alpha = x^\alpha/r^2, \quad r^2 = x^2 + y^2 \qquad (2.94)$$

The plane must be augmented by a 'point at infinity', which is the image of the origin, and whose image is the origin, when (2.94) is applied.

The projective transformations are best expressed in terms of *homogeneous* coordinates χ^i ($i = 1, 2, 3$) for the plane. A projective transformation is a linear homogeneous transformation

$$\chi'^i = A^i_j \chi^j, \qquad (2.95)$$

the 3×3 matrix A^i_j being non-singular. A homogeneous coordinate system can be obtained from a Cartesian one by defining

$$\chi^1 : \chi^2 : \chi^3 = x : y : 1.$$

In this prescription, the 'line at infinity' is the line whose equation is $\chi^3 = 0$.

All of the transformation equations discussed above have two conceptually distinct interpretations. We have been emphasising the *active* interpretation, whereby, with a fixed coordinate system, every point x^α has been assigned an image x'^α. The *passive* interpretation regards x^α and x'^α as coordinates for the same point, in two different coordinate systems. According to this interpretation, the transformation equations specify a change of coordinate system, from an original Cartesian system to a new system which is not Cartesian, but which is specially adapted to the kind of geometry with which we are dealing. Thus, for each geometry, we arrive at the concept of coordinate systems specially adapted to the kind of geometry with which we are dealing. All the transformation laws introduced in this section can thus be re-interpreted (either actively or passively) by regarding both x^α and x'^α as coordinates of such a 'special' system, and dropping all reference to Cartesian systems, which have meaning only in the context of Euclidean geometry.

Chapter 3
Topological spaces

━━━━━━━━━━━━━━━━━━━━━━━━━━━━━━

We have already seen how a Euclidean plane can be regarded as a topological space. We simply disregard all geometrical properties of figures in it except those which remain unchanged under topological transformations [L7, S5]. More picturesquely: we study only those properties of a figure that remain intact when the figure is deformed by arbitrarily distorting the plane as though it were a rubber sheet. In a similar way, any continuous surface can be regarded as a two-dimensional topological space. The topological transformations are one-to-one mappings of the surface onto itself with the property that any continuous curve on the surface has an image which is also a continuous curve.

3.1 Homeomorphisms
If there is a one-to-one correspondence between the points of a surface and those of another surface, so that the topological properties of any figure in one of the surfaces are shared by its image in the other, the two surfaces are said to be *homeomorphic* to each other, and the mapping from one surface to the other established by the one-to-one correspondence is a *homeomorphism*.

Surfaces are two-dimensional topological spaces. Two surfaces that are homeomorphic to each other are topologically indistinguishable. They are to be regarded as two replicas of the *same* topological space. For example, the Euclidean plane and the set of points in the interior of a circle are equivalent as topological spaces. An example of a homeomorphism between them is

$$\left. \begin{array}{l} x' = x(\tanh r)/r \\ y' = y(\tanh r)/r \end{array} \right\} \tag{3.1}$$

The topological space in this example is called a '*disc*'.

3.2 Polygonal Symbols
By introducing a set of cuts, any surface can be made homeomorphic to the Euclidean plane (it can be 'stretched out flat'). In Fig. 8 for example, we see how the surface of a torus can be converted into a rectangle by making two

cuts. The surface can be reconstituted by re-joining the edges that have become separated, paying attention to the correct orientation indicated by the arrows. It is then clear that the surface of a torus, considered only as a *topological space*, is completely described by the *polygonal symbol* [S5]

$$ABA^{-1}B^{-1}$$

which describes the labelling of the perimeter of the rectangle.

As a second example, consider the Möbius strip, which is obtained by putting a twist in a rectangle before joining a pair of opposite edges (Fig. 9). The surface has only one side. A small closed curve on the surface with a sense (clockwise or anticlockwise) attached to it can be moved over the surface until it arrives at its original position with the sense reversed. A surface with this property is said to be *non-orientable*. Surfaces such as the surface of a sphere, or a Euclidean plane, are two-sided or *orientable*. From Fig. 10 it is clear that an appropriate polygonal symbol for the Möbius strip is

$$AKAL$$

By cutting and rejoining the rectangle and *relabelling* (as shown in Fig. 10), we obtain an alternative, simpler, polygonal symbol

$$AAK$$

By joining two Möbius strips AAK and BBK^{-1} along their free edges, as in Fig. 11, we get a *closed* one-sided surface with the symbol

$$AABB$$

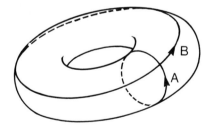
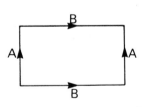

Figure 8

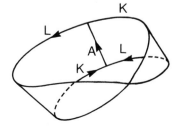
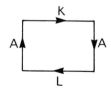

Figure 9

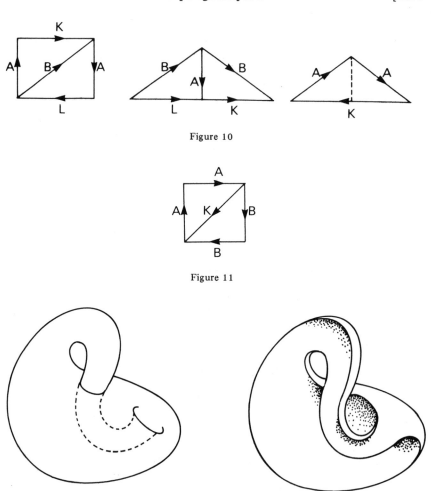

Figure 10

Figure 11

Figure 12

This surface is *Klein's bottle*. It cannot exist in three-dimensional Euclidean space without self-intersection (in four dimensions, it can). A Klein's bottle in 3-space is illustrated in Fig. 12, together with a Möbius strip drawn so as to emphasise its role as a half of Klein's bottle. An alternative symbol for Klein's bottle is

$$CBCB^{-1},$$

as is shown by the cutting and rejoining of the polygon indicated in Fig. 13.

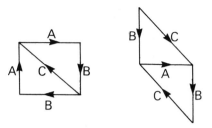

Figure 13

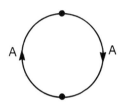

Figure 14

A symbol for the surface of a sphere (or any surface homeomorphic to the surface of a sphere) is

$$AA^{-1}$$

obtained by making a single cut. The disc (i.e. the Euclidean plane considered as a topological space) has a single free edge. No cuts are needed. This surface is therefore symbolised by a single letter, denoting the free edge:

$$K$$

The topological sphere is usually called S_2, and the disc is R_2.

We have seen, in §2.9, that the points of the projective plane are in one-to-one correspondence with the lines through the origin of Euclidean 3-space. By considering the intersections of these lines with a sphere centred at the origin, we obtain a homeomorphism between the projective plane and a sphere with antipodal points identified. Since antipodal points are identified, half of the sphere can be removed, so we get a hemisphere with opposite points on its free edge identified. Stretching this out flat, we get a disc with opposite points of its edge identified (Fig. 14). The symbol for the projective plane, considered only as a topological space, is therefore

$$AA$$

There is no surface that can be embedded in three dimensions without self-intersections that is homeomorphic to this. A surface homeomorphic to the projective plane, but with self-intersections, is shown in Fig. 15.

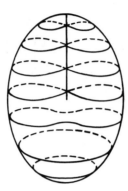

Figure 15

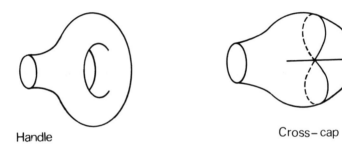

Handle   Cross-cap

Figure 16

Two surfaces of fundamental importance are: the *handle* (a torus with a hole cut in it), and a *cross-cap* (a projective plane with a hole cut in it) (Fig. 16). The introduction of a hole produces a single free edge, which is to be appended to the symbols for the torus and the projective plane. Thus the handle and the cross-cap are symbolised, respectively, by

$$\text{ABA}^{-1}\text{BK}, \quad \text{AAK}$$

It follows that *a cross-cap is homeomorphic to a Möbius strip*.

3.3 Classification of Surfaces
It is now clear that any two dimensional topological space can be specified by a symbol consisting of a string of letters, each occurring at most twice. A letter occurring only once signifies a free edge of the surface (equivalently, a hole in the surface). A letter occurring twice can appear as a direct pair (A and A) or as an inverse pair (A and A^{-1}), corresponding to the two ways of joining a pair of edges of the polygonal representation to reconstitute the surface.

A surface does not have a unique symbol, so there is a need to establish standard canonical forms for the symbols. This will provide a topological classification of all possible surfaces.

We define the *inverse* of a symbol $ABC\ldots$ to be the symbol $\ldots C^{-1}B^{-1}A^{-1}$.

The following properties of symbols are immediately apparent (two symbols are equivalent if they refer to the same surface).

(i) Two symbols differing only in their cyclic order are equivalent. This simply means that, in following the labelling of the perimeter of a polygon, we can begin at any vertex.

(ii) Every symbol is equivalent to its inverse. This means that the perimeter of the polygon can be followed in either direction.

(iii) The actual letters used in a symbol are irrelevant.

(iv) A string of consecutive unpaired letters can be replaced by a single unpaired letter.

(v) If a string of consecutive letters occurs twice in a symbol, it can be replaced by a single letter, directly paired.

(vi) If a string of consecutive letters and its inverse both occur in a symbol, they can be replaced by a single letter, inversely paired.

We shall denote free edges (non-repeated letters) by $K, L, \ldots$ and paired edges by $A, B, \ldots G$. Strings of consecutive letters occurring in a symbol will be denoted by Greek letters.

We consider first the classification of *closed* surfaces. Then every letter in a symbol is paired, either directly or inversely. If α is such a symbol, αK corresponds to the same surface with a hole cut in it. Two closed surfaces α and β can be combined by cutting a hole in each and joining them along the free edges so formed (as we did in the previous section in deriving the Klein bottle from a pair of Möbius strips). Symbolically, $\alpha K + \beta K^{-1} = \alpha\beta$ or $\alpha K + \beta K = \alpha\beta^{-1}$ (Fig. 17).

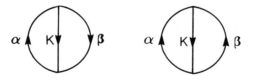

Figure 17

By repeatedly joining handles to a sphere, we deduce the polygonal symbol H_g for a *sphere with g handles*:

$$H_0 = AA^{-1}$$
$$H_g = A_1 B_1 A_1^{-1} B_1^{-1} \ldots A_g B_g A_g^{-1} B_g^{-1} \ (g > 0) \tag{3.3}$$

Similarly, a *sphere with k cross-caps* is symbolised by

$$C_k = A_1^2 A_2^2 \ldots A_k^2 \ (k > 0) \tag{3.3}$$

For example, a torus is a sphere with one handle and a Klein bottle is a sphere with two cross-caps.

Every closed orientable surface is a sphere with g handles (g ⩾ 0) and every closed non-orientable surface is a sphere with k cross-caps (k > 0).

To prove this assertion, we need some additional rules for manipulating polygonal symbols. These are established by cutting and reassembling polygons; we will use the equality sign to denote equivalence of polygonal symbols.

(vii) $A\alpha A\beta = A^2\alpha^{-1}\beta$. This is demonstrated in Fig. 18 (the E in the final symbol $EE\alpha^{-1}\beta$ can be replaced by A according to rule (iii)).

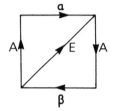 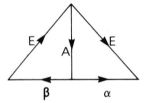

Figure 18

By repeated application of this rule, any symbol can be changed to an equivalent symbol in which all direct pairs appear as *consecutive* pairs of identical letters.

(viii) $C_k\alpha A^2\beta = C_{k+1}\beta^{-1}\alpha^{-1}$. The proof consists of applying the rules (i), (vii), (ii), (i), (vii), (i), in that order (note that $C_k = C_k^{-1}$): $C_k\alpha A^2\beta = A^2\beta C_k\alpha = A\beta^{-1}AC_k\alpha = \alpha^{-1}C_kA^{-1}\beta A^{-1} = A^{-1}\beta A^{-1}\alpha^{-1}C_k = A^{-2}\beta^{-1}\alpha^{-1}C_k = C_kA^{-2}\beta^{-1}\alpha^{-1} = C_{k+1}\beta^{-1}\alpha^{-1}$.

By repeated application of this rule, all direct pairs can be brought to the left-hand end of the polygonal symbol, so that any polygonal symbol is equivalent to one of the forms

$$C_m, \ C_m\pi, \ \pi,$$

where π is a symbol containing only *inverse pairs*.

(ix) $\alpha A\beta B\gamma A^{-1}\delta B^{-1}\epsilon = ABA^{-1}B^{-1}\alpha\delta\gamma\beta\epsilon$. This is demonstrated by the dissection and reassembly illustrated in Fig. 19, followed by the relabelling of E_1E_2 and F_1F_2 by A and B respectively.

Application of this rule to the symbol $C_m\pi$ or π will bring 'handles' to the left of the symbol until we arrive at

$$H_nC_m \quad \text{or} \quad H_m$$

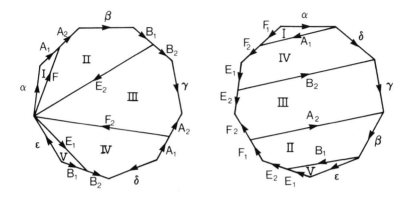

Figure 19

Finally,

(x) $BCB^{-1}C^{-1}A^2\alpha = A^2B^2C^2\alpha$. That is, a handle and a cross-cap occurring together on a surface are equivalent to three cross-caps. To establish this, note that rule (viii) for the special case $k = 0$ is just $\alpha A^2\beta = A^2\beta^{-1}\alpha^{-1}$ which can also be expressed (on account of (i) and (ii)) as $\alpha A^2\beta = \beta A^2\alpha$. Therefore $BCB^{-1}C^{-1}A^2\alpha = \alpha A^2BCB^{-1}C^{-1}$. The proof then follows from the cutting and reassembling shown in Fig. 20, followed by the relabelling of E, F_1 F_2 and G as A, B and C.

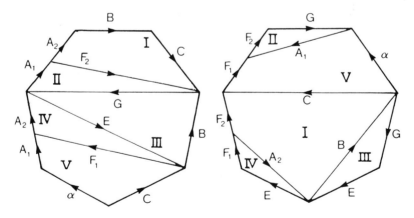

Figure 20

Repeated application of this rule will convert the symbol H_nC_m to C_{n+2m}. We then have the desired result:

The polygonal symbol for any closed surface is equivalent to H_g $(g \geqslant 0)$
or to C_k $(k > 0)$. The number g of handles of an orientable surface is
called the *genus* of the surface.

A slightly neater form than (3.3) for the canonical symbol of a closed
orientable surface can be obtained by applying the rule $ABA^{-1}B^{-1}\alpha\beta =$
$AB\alpha A^{-1}B^{-1}\beta$, established as shown in Fig. 21.

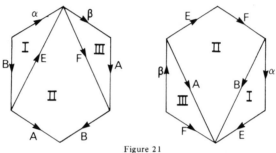

Figure 21

We have, then

$$H_g = A_1 A_2 \ldots A_{2g} A_1^{-1} A_2^{-1} \ldots A_{2g}^{-1}. \qquad (3.4)$$

To obtain the canonical symbols for surfaces with free edges, we note
that the free edges appear as holes in the polygon. They correspond to free
edges $K_1, K_2 \ldots K_n$.

By making the cuts $E_1, \ldots E_n$ as shown in Fig. 22, we obtain the
canonical symbols

$$\left.\begin{array}{l} H_g E_1 K_1 E_1^{-1} \ldots E_n K_n E_n^{-1} \\ C_k E_1 K_1 E_1^{-1} \ldots E_n K_n E_n^{-1} \end{array}\right\} \qquad (3.5)$$

for surfaces with n free edges. The topological classification of surfaces
is then complete.

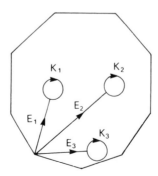

Figure 22

Chapter 4

Mappings

We now arrive at a concept which will serve as a central unifying principle in our approach to the study of mathematical methods of shape description: the idea of a *mapping* from one geometrical space to another. The terminology of mappings was introduced in §2.1. If S and S′ are two spaces, a mapping from S to S′ associates with each point of S a unique image point in S′. Its total action produces a figure in S′; the *image* of S under the given mapping. A mapping from S to S′ will be called *continuous at a point* P of S if every curve in S which is continuous at P has an image curve which is continuous at the image of P. A *continuous* mapping is one that is continuous at every point of S.

4.1 Submersion and Immersion

A continuous mapping from an M-dimensional space S to an N-dimensional space S′ is a *submersion* if N < M and an *immersion* if N > M. If an immersion is one-to-one, then the image of S does not intersect itself (it has no 'double-points') and S will be homeomorphic to its image. The immersion is then called an *embedding* [B12, R6].

Examples. A continuous curve in a space S′ is an immersion of a one-dimensional topological space. There are only two homeomorphically distinct one-dimensional topological spaces, namely R_1 (homeomorphic to a line in Euclidean space) and S_1 (homeomorphic to a circle). Their immersions give images which are, respectively, open curves and closed curves. A surface in Euclidean 3-space is an immersion of a two-dimensional topological space. Fig. 12 is the image of a Klein bottle under an immersion. This immersion is *not* an embedding, because the surface intersects itself. The Klein bottle can be embedded in Euclidean 4-space, but not in 3-space. Fig. 22a, illustrating two embeddings of a torus in 3-space, and Fig. 23b, illustrating two embeddings of a Möbius strip in 3-space, highlight the fact that two spaces may be homeomorphic and yet be embedded in topologically distinct ways. The study of the distinct ways of embedding S_1 in Euclidean 3-space constitutes the theory of knots [D6, L7]. The two simplest embeddings are those shown in Fig. 24. It should be clear

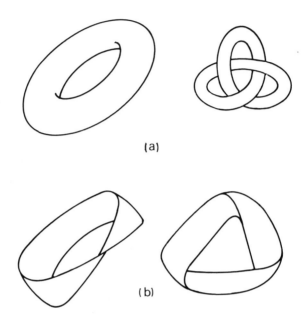

(a)

(b)

Figure 23

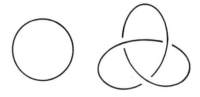

Figure 24

from these few examples, that the mathematical investigation of the notion of 'embedding' in general would be exceedingly complex. It forms a large part of the investigations in pure mathematics that have been pursued in the twentieth century.

4.2 Description of Mappings in Terms of Coordinates

Although, in general, a surface cannot be given a single coordinate system, a piecemeal coordinatisation can be imposed by covering the surface with a set of overlapping patches, each of which is homeomorphic to a disc, and giving a separate coordinate system to each patch (see §2.4). Similarly, an N-dimensional space can be covered by a set of overlapping coordinate patches, each homeo-morphic to N-dimensional Euclidean space.

Let u^α ($\alpha = 1, \ldots M$) be the coordinates for a patch of an M-dimensional space S and let x^i ($i = 1 \ldots N$) be the coordinates for a patch of an N-dimensional space S$'$. A mapping from S to S$'$ can be described by specifying a set of relations

$$x^i = f^i(u^\alpha). \tag{4.1}$$

(the mapping is regarded as built up of a set of mappings from patches of S to patches of S$'$).

We shall call S the *parameter space* and S$'$ the *image space*. A change of coordinates of the image space is described by a set of relations of the form

$$x'^i = a^i(x^j) \tag{4.2}$$

and a change of coordinates of the parameter space is described by a set of relations of the form

$$u'^\alpha = b^\alpha(u^\beta). \tag{4.3}$$

Correspondingly, we have a new description of the mapping:

$$x'^i = f'^i(u'^\alpha) \tag{4.4}$$

If the mapping is continuous, the functions f^i are continuous, the pairs of coordinate systems on an overlap of two patches must be related by continuous functions and allowed coordinate changes (4.2) and (4.3) must be restricted to those expressible with continuous functions a^i and b^α. Analogous stipulations apply to the description of differentiable mappings, and for *analytic* mappings (analytic functions are those which possess valid Taylor expansions about every point at which they are defined).

M = 1, N = 2   (i) Parametric description of a curve in two-dimensional space. Equation (4.3), which is simply $u' = b(u)$, is a change of parameter for the curve.

(ii) If the parameter is time, we have a description of the motion of a point in a two-dimensional space.

M = 1, N = 3   (i) Curves in 3-space.

(ii) Motion of a point in 3-space. More generally, the time-development of a dynamical system with N degrees of freedom can be specified by the motion of a point in an N-dimensional space.

M = 2, N = 1   A *scalar field* in two dimensions. The number $x = f(u, v)$ is the value of the field at the point (u, v). A visualisation of a scalar field in two dimensions is obtained by regarding the equation as the equation of a surface, in Monge's form, in the three-dimensional space with coordinates (u, v, x) (Fig. 25a).

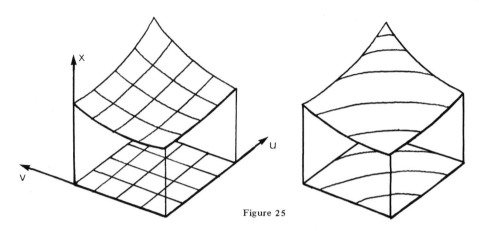

Figure 25

Another useful visualisation is in terms of *contours*, which are the curves x = const. in (u, v)-space (Fig. 25b).

$M = 2$, $N = 2$ (i) A *mapping between two surfaces*. In cartography, a basic set of problems is concerned with the mapping of a portion of a sphere on a Euclidean plane (see §4.3).

(ii) *Deformation of a surface*. If a coordinate system is attached to a surface so that a point retains the same coordinate pair when the shape of the surface changes, equation (4.1) is simply $x^\alpha = u^\alpha$ ($\alpha = 1, 2$). The change of shape is then described by the changes in the components of the metric, and, if the surface is immersed in 3-space, by the changes in the components of the second fundamental form.

(iii) If one of the parameters is *time*, we get a description of the time-development of a moving curve, in a two-dimensional space. More generally, $x^\alpha = f^\alpha (u, v)$ describes a *family of curves* if the transformations (4.3) are restricted to transformations of the form

$$\left. \begin{array}{l} u' = b\,(u, v) \\ v' = c\,(v) \end{array} \right\} \qquad (4.5)$$

For fixed v, $x^\alpha = f^\alpha (u, v)$ gives a parametric description of one curve of the family.

(iv) *Deformation of an object* in a two-dimensional space (see §4.5). Here, $x^\alpha = f^\alpha (u^\beta)$ gives the coordinates of the new position of the point in the deformed state that originally had the coordinates u^α. We are dealing here with only one coordinate system, so the functions a^α and b^α of (4.2) and (4.3) are the same.

(v) A *vector field* in a two-dimensional space. Here, $x^\alpha = f^\alpha (u^\beta)$ are the components, referred to the coordinate system u^α, of the vector at the point with coordinates u^α. With this interpretation, (4.2) and (4.3) are restricted to the form

$$\left. \begin{array}{l} x'^\alpha = \dfrac{\partial b^\alpha}{\partial u^\beta} \, x^\beta \\[2mm] u'^\alpha = b^\alpha (u^\beta) \end{array} \right\} \qquad (4.6)$$

M = 2, N = 3 (i) Parametric description of a *surface* in 3-space.

(ii) A *family of curves* in 3-space, or the time-development of a curve in 3-space.

M = 3, N = 1 (i) A *scalar field* in 3-space.

(ii) Time-development of a scalar field in 2-space.

M = 3, N = 2 (i) Time-development of the continuous deformation of an object, in 2-space.

(ii) Time-dependent deformation of a surface.

M = 3, N = 3 (i) Deformation of an object in 3-space (see § 4.5).

(ii) A family of surfaces in 3-space, or the time-dependent deformation of a surface in 3-space.

(iii) *A vector field in 3-space.*

M = 4, N = 1 (i) Time-development of a scalar field in 3-space.

M = 4, N = 3 (i) Time-development of a vector field in 3-space.

(ii) Time dependent deformation of an object in 3-space (see §4.8).

4.3 The Design of Boat Hulls

An interesting application of the use of contours for the description of the shape of a surface is the representation, on a drawing, of the shape of the hull of a boat [O3]. The *'sections'* are the intersections of the hull by vertical planes perpendicular to the main axis of the boat (Fig. 26a). The sections in the fore and aft parts of the boat are represented separately on the right and the left halves of the figure. The *bow and buttock lines* are intersections of the hull by vertical planes running parallel to the main axis (Fig. 26b) and the *water lines* are intersections by horizontal planes (Fig. 26c).

The diagrams are taken from Peter Heaton's *Sailing* [H6], and represent the hull of the 'Nicholson 26' designed by Charles Nicholson.

Three sets of contours of a surface, corresponding to three mutually perpendicular sets of planes, cannot, of course, be specified independently. The shapes of the three sets of contours have to be designed simultaneously, by a process of trial and error, known as 'fairing'. A computer-assisted method of fairing developed by Duncan and Vickers [D8] will be discussed in § 10.3.

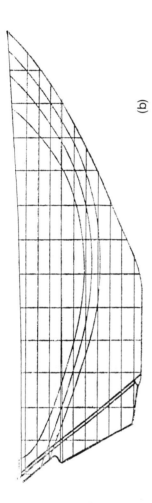

(b)

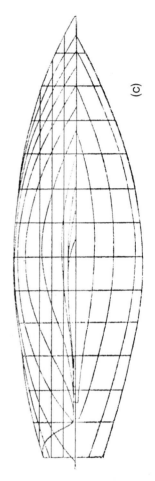

(c)

Figure 26

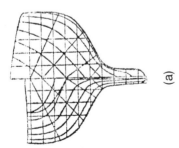

(a)

4.4 Map Projections

The production of a map of a large area of the earth's surface requires the establishment of a homeomorphism between a portion of the surface of a sphere and a portion of the Euclidean plane [H8, S15, W2]. The choice of the homeomorphism is influenced by the intended use of the proposed map. For example, the purpose of a map may require that it shall have one of the following properties:

(i) Correct representation of angles (and, consequently, of the shape of *small* regions). *Conformal* projections.
(ii) Correct representation of ratios of areas. *Equal-area* projections.
(iii) Correct representation of bearings from some chosen point. *Azimuthal* or *zenithal* projections.

Of course, it is not possible for a single map to have all three properties, and there may be other requirements in addition to these. Every representation of the earth's surface on a plane involves compromise. The study of map projections is thus an excellent example of the issues involved in the problem of choosing a method of shape description, on the basis of the purpose for which the description is required, and of the ingenuity in the resolution of such a problem.

The most obvious way of establishing a homeomorphism between a portion of a sphere and a portion of a plane is by projection from a fixed point. This operation, with three different choices for the position of the fixed point, is illustrated in Fig. 27. The choices correspond to *gnomonic, stereographic,* and *orthographic* projection. All three projections preserve bearings from a point P — they are azimuthal. Gnomonic projection maps all great circles to straight lines, hence its usefulness for indicating air-routes. The region displayed by the map is necessarily less than a hemisphere, because gnomonic projection maps a hemisphere onto the whole plane. However, an ingeneous method of obtaining a world map is by gnomonic projection of the sphere onto a circumscribing polyhedron. We get a set of polygonal-shaped maps that match along their edges. Gnomonic projection onto a cube is discussed by Hinks [H10] and Steers [S15].

As well as being conformal, stereographic projection also has the property that every circle on the sphere is represented by a circle (or a straight line) on the plane.

Orthographic projection represents a hemisphere by the interior of a circle. Such a map corresponds to the view of the earth from a great distance. The degree of distortion near the circumference of the circle is extreme, so the orthographic projection is not often used.

The central mathematical problem of map projection is the determination of the images of the lines of latitude and longitude (parallels and meridians). The three projections of Fig. 27 are further classified as *normal, oblique,* and *equatorial,* according to the position of the north-south axis with respect to the position of the plane and point of projection (Fig. 28). The appearance of the

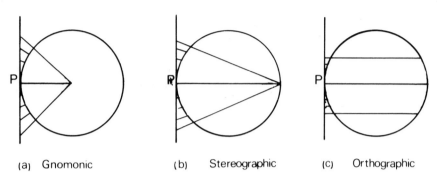

(a)   Gnomonic   (b)   Stereographic   (c)   Orthographic

Figure 27

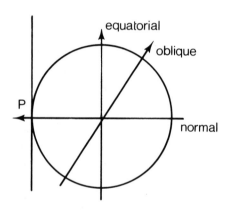

Figure 28

images of the parallels and meridians, in these three cases, are indicated for the stereographic projection in Fig. 29 (the dotted line is the image of the great circle parallel to the plane of projection).

An azimuthal equal-area projection is obtained if the equator is represented by a circle, and one of the poles by its centre. The meridians are represented by radii, so that bearings from the pole are correctly represented. The parallels are represented by concentric circles whose radii are computed so as to achieve the equal-area property (taking the radius of the equatorial circle to be 1, the radius of the circle representing co-latitude ζ will be $\sin \frac{1}{2} \zeta$). The result is the normal form of *Lambert's* azimuthal equal-area projection, commonly used in atlases to display the polar regions. The image of a hemisphere in the transverse version of Lambert's projection is shown in Fig. 30. If this figure is stretched uniformly by a factor 2 in the direction of the equator, we obtain an elliptical figure. Relabelling the longitude lines in an obvious way provides a basis for an equal-area projection of the *whole* surface of the sphere. This is *Aitoff's projection* (Fig. 31).

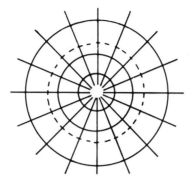

(a) Normal stereographic

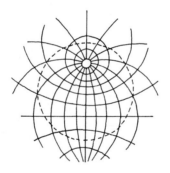

(b) Oblique stereographic

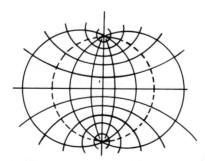

(c) Equatorial (or transverse) stereographic

Figure 29

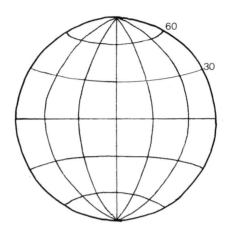

Figure 30

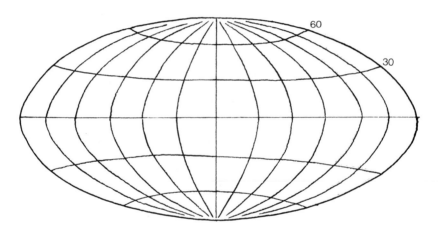

Figure 31

The requirement for a general-purpose chart of a large portion of the earth's surface, such as those found in atlases, is that the amount of distortion should be kept as small as possible over a large area. The various kinds of *conical* projection, and their modifications, are designed to satisfy this requirement. The basic idea behind conical projection is the establishment of a mapping of a portion of a sphere onto a portion of a cone. Since a cone has the same intrinsic geometry as the Euclidean plane, it can be cut and laid out flat on a plane.

In the *normal conical projection with one standard parallel*, a cone is imagined to touch the sphere along one line of latitude. The meridians are represented as generators of the cone. The mapping is specified by requiring that distances along the standard parallel, and distances along the meridians,

are correctly represented. This causes the north and south poles to be represented by circles (Fig. 32a); the chart is accurate close to the standard parallel
but the degree of distortion (stretching of lines of latitude) increases as we move
away from the standard parallel. Fig. 32b shows the image of the whole sphere,
when the cone is laid out flat, with lines of latitude and longitude marked. Of
course, only a portion of this, with the standard parallel running across near the
middle, would be used as a map.

The *conical projection with two standard parallels* is a generalisation which
keeps the distortion small over a larger region. The general appearance of the
image of the whole sphere is much like Fig. 32b, but distances are correctly
represented along the meridians and along *two* chosen parallels. The distortion
consists of stretching of the latitude lines that lie outside the region between the
standard parallels, and a contraction of those within the region.

If the standard parallel for the simple conical projection is chosen to be
the equator, we get a *cylindrical* projection. The image of the whole earth is a
double square, with images of the parallels and meridians dividing it into a grid
of small squares. If two standard parallels are chosen at equal latitudes north and
south of the equator, a cylindrical projection with two standard parallels is
obtained.

A very popular projection used in atlases is *Bonne's projection*. This is a
modification of the conical projection with one standard parallel. The circles
of latitude, and *one* meridian, are represented as in Fig. 32b, but the remaining

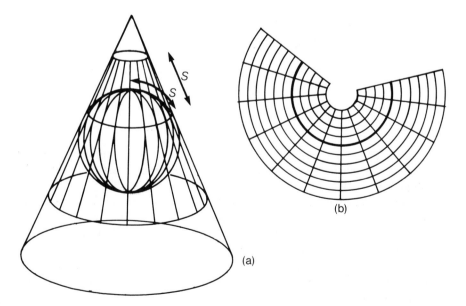

Figure 32

meridians are represented as curves so as to give correct representation of distances along all the parallels, not just the standard parallel. Bonne's projection is an equal-area projection. Minimal distortion over a large area is achieved by choosing the point of intersection of the standard parallel and standard meridian near the centre of the region to be mapped. Bonne's projection can also be based on two standard parallels. The appearance of the world map in Bonne's projection (which would not, of course, be used in practice) is indicated in Fig. 33. The rectangle indicates a typical map area within which the distortion is within 'reasonable' limits.

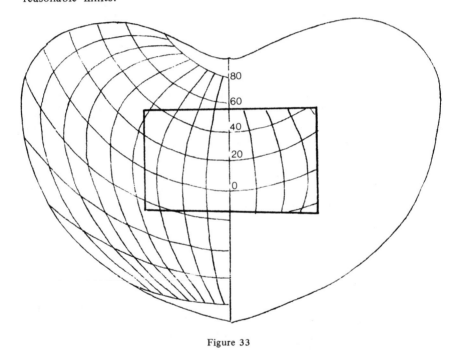

Figure 33

Sanson-Flamsteed's projection is the special case of Bonne's projection, with the standard parallel chosen to be the equator. The meridians are represented by sinusoidal curves (Fig. 34).

A very familiar-world map is *Mollweide's projection*, an equal-area mapping of the sphere onto the interior of an ellipse whose major axis is twice the minor axis. In the normal version of Mollweide's projection (Fig. 35), the major axis is the equator and the minor axis a chosen meridian. Distances are correctly represented along the equator and the meridians are half-ellipses. The spacing of the lines parallel to the major axis, that represent lines of latitude, is determined by the equal-area requirement (if the semi-minor axis is 1, latitude ϕ is represented by a line a distance $\sin \theta$ from the major axis, where $2\theta + \sin 2\theta = \pi \sin \phi$).

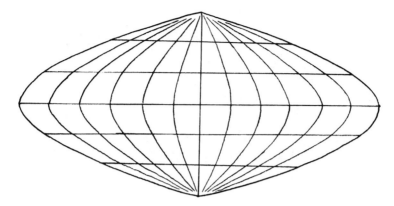

Figure 34

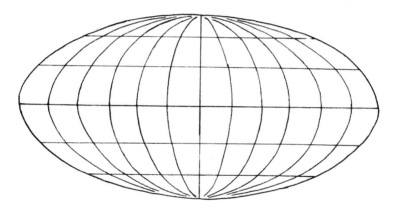

Figure 35

Mollweide's projection is more common than Aitoff's (which is also equal-area and represents the world as an ellipse) although the degree of distortion near the poles is more severe than in Aitoff's projection.

An interesting example of the way the choice of a method of form description can be influenced by the nature of the object whose form is described, is provided by Mollweide's interrupted projection. In this projection, the surface of the earth is mapped onto the interior of a shape built up of semi-ellipses (Fig. 36). The 'interruptions' are chosen to lie in the oceans. This achieves a degree of distortion of the shapes of the land-masses that is much smaller than in the basic Mollweide projection.

A very common and useful projection is *Mercator's*, in which the meridians are represented by equally spaced parallel lines, and the latitude lines are orthogonal to them, their spacing being computed so as to achieve the *conformal*

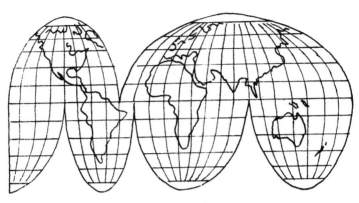

Figure 36

(equal-angle) property. This projection is used for navigational purposes, because bearings at any point are correctly represented. The rhumb-lines or loxodromes (lines of constant bearing — Fig. 37) are represented by straight lines. As a general-purpose world map it is unsuitable (though quite common) because the ratios of areas near the equator to those at high latitudes are completely falsified.

All the map projections we have discussed may be normal, oblique, or transverse, but some forms are very rarely used. A transverse version of Mollweide's projection is shown in Fig. 38 [H10, S15].

The number of different map projections that have been suggested and employed is very large, and their study is a fascinating branch of geometry.

This chapter would be incomplete without a mention of Buckminster Fuller's remarkable polyhedral projections [M7]. We shall describe the one based on the cuboctahedron. Fig. 39 illustrates a cuboctahedron and the gnomonic projection *of its edges* on a sphere. The equal division of the edges in both

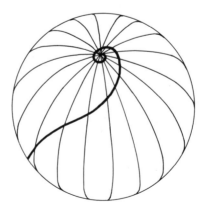

Figure 37

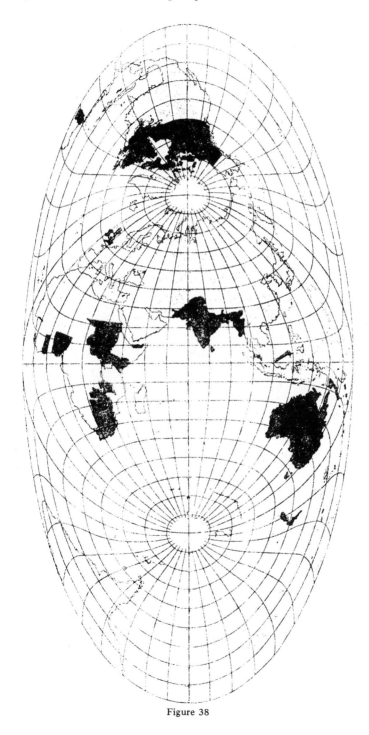

Figure 38

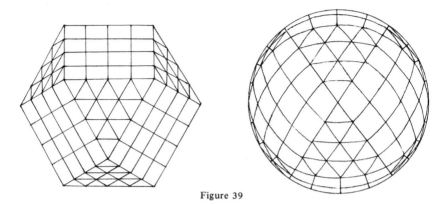

Figure 39

cases, and the corresponding subdivision of the faces (by straight lines in the case of the cuboctahedron and by arcs of great circles in the case of the sphere) determines a homeomorphism between the two surfaces. It is *not* gnomonic. We obtain a set of maps on the faces of the cuboctahedron, with *distances correctly represented on the edges*. They can be assembled on a plane to form a world map (Fig. 40). Some ingenuity is required in choosing the position of the distribution of continents relative to the octahedron, and the method of assembling the component maps on the plane, in order to get a good distribution of land areas on the world map. The degree of distortion is considerably less than that of the map obtained by gnomonic projection onto the cuboctahedron.

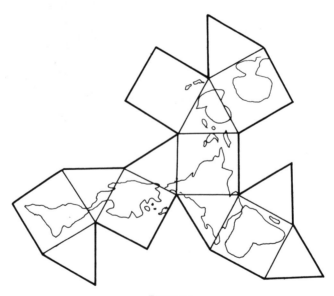

Figure 40

4.5 Illumination Engineering

The orthographic projection (Fig. 27c) finds a practical application in the estimation of the illumination received at a point from a set of light-sources. We illustrate the method by a simple example. Assume that an overcast sky can be regarded as a hemisphere with uniform luminosity. Let P be a point in a room with windows and roof-lights. Take P at the centre of a hemisphere. The contributions to the scalar illumination at P due to various openings are proportional to the areas of the images on the hemisphere of the various openings, produced by projection from P (Fig. 41).

The contributions to the *horizontal* illumination at P, due to the various openings, are proportional to the areas of the images of the spherical areas, under orthographic projection of the hemisphere on the horizontal plane. (Fig. 42). In this orthographic image, all horizontal lines of the room in which P lies are represented as elliptical arcs and all vertical lines by radii of the circle. A more detailed discussion of the method, and its extension to deal with point sources and vertical illumination, is given in Lynes' *Principles of Natural Lighting* [L11].

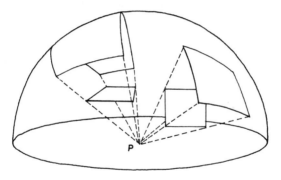

Figure 41

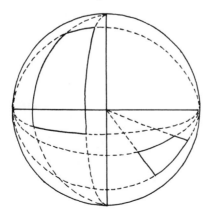

Figure 42

4.6 Deformations in Euclidean Space

If an elastic solid is subjected to stresses, a deformation will be produced, and a means of describing the deformation is required [L2]. What we are dealing with, in all problems of this kind, is a homeomorphic mapping of a region of Euclidean space onto another.

Let (x^1, x^2, x^3) be the Cartesian coordinates of a point in the undeformed material, and let $(\bar{x}^1, \bar{x}^2, \bar{x}^3)$ be the coordinates, in the *same* coordinate system, of the same material point, after the deformation. The deformation can then be described by a *displacement vector field* ζ, whose components at the point x^i are $\zeta^i = \bar{x}^i - x^i$. The analogous two-dimensional situation is shown in Fig. 43.

Alternatively, the deformation is described by the functions f^i ($i = 1, 2, 3$), where

$$\bar{x}^i = f^i(x^j) \tag{4.7}$$

These functions are single-valued, and if the deformation does not involve any disruption of the material, they will be continuous functions. Also, if the bending of the material takes place 'smoothly', they will be differentiable. The single-valuedness is in that case expressed by the condition that the Jacobian $|\partial f^i/\partial x^j|$ is nowhere zero in the region to be deformed. Since the coordinate system x^i is Cartesian, the distance ds between two neighbouring points is given by

$$ds^2 = \sum_{i=1}^{3} (dx^i)^2 \tag{4.8}$$

before the deformation. After the deformation, the distance between the same points of the deformed substance will be given by

$$d\bar{s}^2 = \sum_{i=1}^{3} (d\bar{x}^i)^2 = g_{ij} dx^i dx^j \tag{4.9}$$

where

$$g_{ij} = \sum_{k=1}^{3} \frac{\partial f^k}{\partial x^i} \frac{\partial f^k}{\partial x^j} \tag{4.10}$$

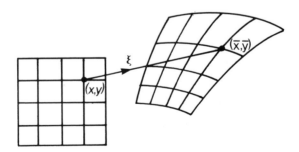

Figure 43

The Cartesian system is called the *Lagrangian* system. The curvilinear co-ordinate system obtained by relabelling the material points of the deformed object with their *old* coordinate labels is called the *Eulerian* system. The co-ordinate transformation from Lagrangian coordinates x^i to Eulerian coordinates x'^i satisfies

$$\frac{\partial x^i}{\partial x'^j} = \frac{\partial f^i}{\partial x^j},\tag{4.11}$$

from which it follows that the g_{ij} defined by (4.10) are the components of the metric of the Euclidean space, expressed in terms of the Eulerian system. The curvature tensor associated with these quantities is therefore zero (see §2.8).

If we are interested only in the change of form produced by the deforma-tion, and not in the accompanying change in position and orientation of the object, the displacement vector contains irrelevant information. *All the infor-mation about the change of form is contained in the fields g_{ij}.*

A fascinating application of the use of Lagrangian and Eulerian coordinate systems, as a means of describing a deformation, is d'Arcy Thompson's method of comparing the shapes of two related organisms by regarding one as a *defor-mation* of the other (Fig. 44) [T3]. Thompson employed this means of shape description to studies of sequences of evolutionary changes in the shapes of bones. The successive changes of the Eulerian coordinate net in such sequences is quite strikingly systematic, so that 'missing links' in the chain can be inter-polated.

In the foregoing discussion, the initial coordinate system was taken to be Cartesian. This is, of course, not necessary (and, in the two-dimensional case of deformation of a surface that is not initially flat, it will not be possible). If

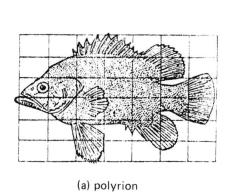

(a) polyrion

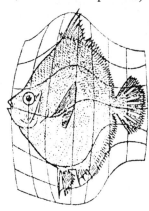

(b) Antigonia capros

Figure 44

a curvilinear coordinate system (x^1, x^2, x^3) is deformed along with the material (so that a material point will have the same set of coordinates before and after the deformation), the distance between two neighbouring points, before and after the deformation respectively, will be given by

$$\left. \begin{array}{l} ds^2 = g_{ij}\, dx^i dx^j \\ \overline{ds}^{\,2} = \overline{g}_{ij}\, dx^i dx^j \end{array} \right\} \tag{4.12}$$

The deformation alters the distance between the points by a factor $\lambda = \overline{ds}/ds$.

Let a direction at a point in the undeformed material be specified by a until vector γ^i,

$$g_{ij}\, \gamma^i \gamma^j = 1 \tag{4.13}$$

Then, the expansion factor λ along this direction, produced by the deformation, is given by

$$\overline{g}_{ij}\, \gamma^i \gamma^j = \lambda^2 , \tag{4.14}$$

The *principal values* of the expansion factor are given by the stationary values of λ^2 as γ^i is varied subject to the constraint (4.13). The principal values of λ^2 are found to be the *eigenvalues* of the matrix

$$\lambda^i_j = g^{ik}\overline{g}_{kj} \tag{4.15}$$

The corresponding *principal direction* or *principal axes* of the deformation are the eigenvectors. They are orthogonal to each other.

The situation can be visualised by imagining an 'infinitely small' sphere centred at the material point in question. After the deformation, it will have become an ellipsoid (the *indicatrix* of the deformation). The principal directions are the principal axes of the ellipsoid and the principal values are the ratios of their lengths to the diameter of the original sphere. There are three mutually orthogonal families of surfaces whose intersections in the neighbourhood of any point are in the principal directions at that point. If the coordinate system is chosen so that these three families are $x = x^1 = $ const., $y = x^2 = $ const. and $z = x^3 = $ const., the metric will be diagonal before and after the deformation.

$$\left. \begin{array}{l} ds^2 = A\, dx^2 + B\, dy^2 + C\, dz^2 \\ \overline{ds}^{\,2} = A\lambda_1^2\, dx^2 + B\lambda_2^2\, dy^2 + C\lambda_3^2\, dz^2 \end{array} \right\} \tag{4.16}$$

The two-dimensional case applies to the estimation of the degree of distortion inherent in a map projection [W2], which is a deformation of a portion of a sphere, the deformed state being a portion of a plane. For a conformal projection, the indicatrices are circles (whose radii indicate the expansion factor at each point); the principal directions are undefined, since λ^α_β is a multiple of the unit matrix. For an equal-area projection, the indicatrices are ellipses whose areas are equal to the areas of the circles from which they are obtained; the principal values of the deformation satisfy $\lambda_1 \lambda_2 = 1$.

4.7 Conformal Mappings in the Euclidean Plane

A continuous differentiable mapping of a region of a Euclidean plane onto another can be described by two functions

$$\left.\begin{array}{l} \overline{x} = \phi(x, y) \\ \overline{y} = \psi(x, y) \end{array}\right\} \tag{4.17}$$

where x and y are Cartesian coordinates and $\overline{x}$ and $\overline{y}$ the Eulerian coordinates of the deformation (cf. (4.7)). A necessary and sufficient set of conditions for the mapping to be conformal (angle-preserving) is

$$\left.\begin{array}{l} \phi_x = \psi_y \\ \phi_y = -\psi_x \end{array}\right\} \tag{4.18}$$

The subscripts denote differentiation (the alternative condition $\phi_x = -\psi_y$, $\phi_y = \psi_x$ correspond to reversal of the sense of all angles, their magnitudes being preserved).

Now these are precisely the *Cauchy-Riemann* conditions [P7] — the necessary and sufficient conditions for the function $w = \phi + i\psi$ of the complex variable $z = x + iy$ to be single-valued and differentiable. A complex function satisfying these conditions within a region is said to be *regular* in the region. The study of conformal mappings in the plane is therefore the same as the study of regular complex functions.

For a steady, irrotational two-dimensional flow-pattern of a non-viscous incompressible fluid, the velocity potential ϕ and stream function ψ satisfy the conditions (4.18) [M9]. Hence these very restricted, idealised flow patterns are describable by regular complex functions $w(z)$. Fig. 45 shows the streamlines and equipotentials of the flow specified by

$$w = \log \frac{z + a}{z - a} \tag{4.19}$$

It corresponds to the flow due to a source and a sink at the points $x = \pm a$ on the x-axis. Any conformal deformation of a flow pattern will lead to another possible flow pattern. This just corresponds to the fact that a regular function of a regular function is regular in the original complex argument.

Electrostatics in two dimensions is mathematically equivalent to this kind of fluid flow, electrostatic potential corresponding to velocity potential and lines of force corresponding to streamlines. Thus Fig. 45 also represents the lines of force and equipotentials due to a positive and a negative charge (actually lines of charge perpendicular to the plane).

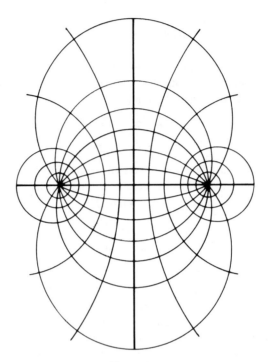

Figure 45

The function

$$w = z + \tfrac{1}{2} \qquad\qquad (4.20)$$

specifies the flow pattern shown in Fig. 46, corresponding to flow round a circular cylinder. The cylinder can be deformed by a conformal mapping, and hence the pattern of flow around a cylinder whose section is *not* circular can be derived from this case. For example, a circle through $z = -c$ enclosing the point $z = c$ and cutting the line joining these points at a small distance from c, will be deformed to an aerofoil-like shape by *Joukowski's transformation*

$$\frac{w - kc}{w + kc} = \left(\frac{z - c}{z + c}\right)^k \qquad\qquad (4.21)$$

The case $c = 1$, $k = 2$ is illustrated in Fig. 47. The angle at the trailing edge of the aerofoil is $(2 - k)\pi$. Of course, the flow pattern around an aerofoil obtained by this method gives only a very crude first approximation, which has to be modified, by perturbation methods, to fit a realistic aerofoil section and to take account of viscosity.

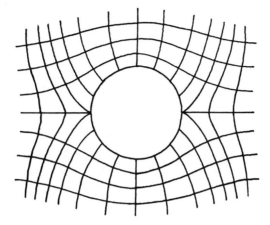

Figure 46

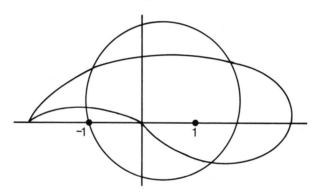

Figure 47

An important set of conformal mappings are the Schwarz-Christoffel trans-formations [M9, P7] , which will map the upper half-plane ($y > 0$) to the exterior of any polygon. In principle, it is possible to convert any closed curve to any other, by a conformal mapping. A catalogue of the more important conformal transformations has been compiled by Kober [K6] .

Unfortunately, the method of conformal mapping cannot be extended to deal with three-dimensional flow patterns (except for problems with cylindrical symmetry), because a conformal mapping in three dimensions is a sphere-pre-serving transformation. Thus, the group of conformal mappings in 3-space is very much more restricted. In the Euclidean plane, the circle-preserving trans-formations, of which Fig. 45 is an example, are just special cases of conformal mappings. Moreover, while the Laplace equation (satisfied by ϕ and ψ) is con-formally invariant, the analogous three-dimensional Laplace equation is not.

4.8 Infinitesimal Deformations

For an infinitesimal deformation, the quantities e_{ij} defined by

$$\bar{g}_{ij} = g_{ij} + e_{ij} \tag{4.22}$$

are the components of the infinitesimal *strain tensor*. If the original coordinate system is Cartesian (so that $g_{ij} = \delta_{ij}$, the components of the unit matrix), they can be expressed in terms of the displacement vector,

$$\overset{\circ}{e}_{ij} = -\partial_i \zeta_j - \partial_j \zeta_i \tag{4.23}$$

The fact that only *derivatives* of the displacement vector occur in (4.23) indicates that translation of an object in space is not a deformation, and the absence of the curl of the vector $(\partial_i \zeta_j - \partial_j \zeta_i)$ corresponds to the fact that rotation of an object in space is not a deformation. Because of (4.23), the strain tensor satisfies the identity

$$\partial_{il}\overset{\circ}{e}_{jk} - \partial_{jl}\overset{\circ}{e}_{ik} + \partial_{jk}\overset{\circ}{e}_{il} - \partial_{ik}\overset{\circ}{e}_{jl} = 0 \tag{4.24}$$

This is just the infinitesimal expression of the fact that, since $\bar{g}_{ij}$ is a metric for Euclidean space, its curvature tensor vanishes (see §2.7). Equation (4.29) is the *Saint-Venant* condition. It is a necessary and sufficient condition for a symmetrical matrix $\overset{\circ}{e}_{ij}$ to have the form (4.23) and hence to be capable of interpretation as an infinitesimal strain. It can be thought of as the requirement that the small regions of a material shall fit together consistently after the deformation. If the straining of a material body violates the condition, cracking and splitting will necessarily occur. An example of a deformation violating the condition occurs in drying mud, where the faster rate of drying at the surface, and consequent faster shrinkage at the surface, can be described by an infinitesimal strain tensor whose differential properties violate the Saint-Venant condition; the cracking of the surface is the result.

The deformation of a two-dimensional flat surface (e.g. a flat plate or membrane) need not satisfy the Saint-Venant condition, because a two-dimensional object need not remain Euclidean. It is capable of *buckling*, or, more precisely, of developing an intrinsic geometry that is not Euclidean. We then have

$$\bar{g}_{\alpha\beta} = g_{\alpha\beta} + e_{\alpha\beta} \ (\alpha, \beta = 1, 2) \tag{4.25}$$

where $g_{\alpha\beta}$ and $\bar{g}_{\alpha\beta}$ are the metrics describing the intrinsic geometry of a surface before and after an infinitesimal deformation, and the differential properties of the infinitesimal strain tensor $e_{\alpha\beta}$ are unrestricted.

4.9 Time-Dependent Deformations

The need to consider continuous time-dependent changes of form is encountered in fluid dynamics [L1, M9]. The growth of an organism is another example of time-dependent deformation, differences in growth rate of different parts of the

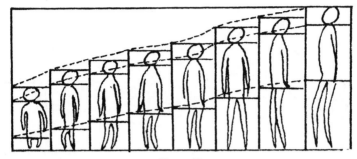

Figure 48

organism producing a change of proportion accompanying the overall increase in size (Fig. 48).

The problem of describing changes in proportion of a growing organism is confronted for the first time by Julian Huxley in 1932, in his *Problems of Relative Growth* [H18]. Unhindered rapid growth is characterised by exponential dependence of dimensions on time. In many organisms growth slows and eventually ceases as the organism becomes adult. Hence a simplified model of the relative growth rate of two parts of an organism, with characteristic lengths A and B respectively, is provided by the equations

$$\left.\begin{aligned} \dot{A} &= \alpha AG \\ \dot{B} &= \beta BG \end{aligned}\right\} \tag{4.26}$$

where α and β are constants and G is a time-dependent factor characteristic of the whole organism (and falling to zero as the organism approaches its maximum size). This implies

$$\frac{d}{dt}(\log A) = k\,\frac{d}{dt}(\log B) \tag{4.27}$$

where $k = \alpha/\beta$. *The ratio of relative growth rates of the two parts is constant.* This law, or its equivalent expression

$$A = bB^k, \tag{4.28}$$

is Huxley's *law of allometric growth*. The law was found by observation to be approximately satisfied in a large number of instances. Of course, the formula can be expected only to give a crude and inaccurate picture of the phenomenon of differential growth, since the growth constant α (or β) is not uniform throughout an organ. See Waddington [W1] for an interesting elementary discussion of the concept of allometry.

The variation of a growth constant α throughout an organism leads to the concept of a *growth gradient*. We illustrate this with a simple one-dimensional example (Fig. 49).

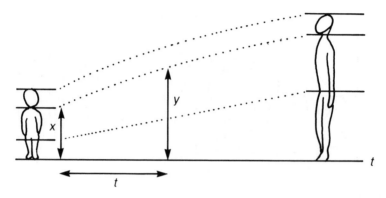

Figure 49

Suppose equation (4.26) is valid for every small portion of height of an organism. Let x be the height of a part of the anatomy at some initial time, and let y be its corresponding height at a later time. Then

$$\frac{d}{dt}(dy) = \alpha(y)G(t)dy \qquad (4.29)$$

The changes in height and corresponding changes in proportion as the organism grows are governed, in this model, by a function $G(t)$ describing the general conditions of growth affected by age and environment, and a 'growth potential' $\alpha(x)$. The solution of (4.29) is

$$y = \int_0^x e^{\alpha(x)\tau}dx, \quad \tau = \int_0^t G(t)dt \qquad (4.30)$$

In the three-dimensional case, a time-dependent deformation can be described by a displacement vector ζ which depends on time as well as on a set of Cartesian (Lagrangian) coordinates. The *velocity field* is

$$u = \dot{\zeta} \qquad (4.31)$$

Corresponding to (4.23), we can define a *rate of strain tensor*

$$\overset{\circ}{\sigma}_{ij} = -\partial_i u_j - \partial_j u_i \qquad (4.32)$$

Now introduce a curvilinear coordinate system x^i, fixed to the material in the sense that the coordinate curves are deformed as the material is deformed so that a material point retains the same set of coordinates. Then if g_{ij} is the metric of this changing system,

$$\dot{g}_{ij} = \sigma_{ij}$$

where σ_{ij} are the components of the rate of strain tensor referred to the curvilinear system. The relative growth rate of the line between two neighbouring points is

$$\sigma = \frac{d}{dt}(ds)/ds$$

The *principal relative growth rates* at a point are the stationary values of σ as the direction is varied. It is easily shown that they are the eigenvalues of

$$\tfrac{1}{2}\sigma_j^i = \tfrac{1}{2}g^{ik}\sigma_{kj} \tag{4.33}$$

and the corresponding *principal directions of growth* (or *axes of growth*) are the eigenvectors (which are mutually orthogonal). An obvious generalisation of (4.29) is the hypothesis

$$\sigma_j^i = \alpha_j^i(x^k)\,G(t) \tag{4.34}$$

The principal axes specify three mutually orthogonal families of surfaces. The hypothesis (4.34) implies that these surfaces are deformed as the material is deformed, as though attached to it. If they are chosen as the coordinate surfaces, the metric will be diagonal and will remain diagonal. In terms of this coordinate system, the development of the organism would be described by

$$ds^2 = e^{2\,\alpha\tau}dx^2 + e^{2\,\beta\tau}dy^2 + e^{2\,\gamma\tau}dz^2$$

$$\tau = \int_0^t G(t)\,dt \tag{4.35}$$

where α, β and γ are the eigenvalues of the matrix α_j^i appearing in (4.34). They are, in general, dependent on position.

These simple mathematical models of organic growth should serve to emphasise that the methods of differential geometry will be an essential tool for acquiring a better understanding of the very difficult problems of explaining the development of biological forms. The central problem of morphogenesis in biology is the identification of the mechanisms that determine the principal axes of growth and the principal relative growth rates in an organism. In his book *Sensitive Chaos*, Schwenk [S4] has pointed out many similarities between the forms of biological systems and the forms characteristic of fluid flow; this suggests that the underlying mathematical equations governing growth may be similar to those governing the motions of a viscous fluid. However, there are some essential differences. In the first place, the velocities involved in growth are so small that inertial forces will be entirely negligible. In the second place, the continuity equation, which expresses the fact that fluid is not created or destroyed, will not apply, since it is the essential characteristic of growth that new material is produced from the supply of nutrients. Finally, the boundary conditions that determine the form of a particular fluid flow will be replaced by *internal* conditions governing changes in growth ratio and growth axes.

Chapter 5

Singularities of analytic mappings

In §4.2, we discussed the way in which a mapping from one space (the parameter space) to another (the image space) gives rise to a description of a form in the image space. The description consists of a set of functions

$$x^i = f^i(u^\alpha). \tag{5.1}$$

The *analytic* functions are particularly 'well-behaved' mathematical functions. They are continuous, differentiable any number of times, and can be expanded in a Taylor series about any point (of parameter space) at which they are defined. We are thus led to the consideration of the class of forms describable in terms of analytic functions, through relations of the kind (5.1).

In this chapter, we shall investigate some of the peculiar form characteristics that can arise in forms describable by analytic functions, namely, the qualitative nature of a form in the neighbourhood of a *singularity*, or *critical point*, of its description.

A *critical point* of a mapping (5.1) is a point in parameter space at which the matrix $\partial f^i/\partial u^\alpha$ has less than maximal rank. All other points are called *regular*.

The general theory of critical points is dealt with in Eells's *Singularities of Smooth Maps* [E2] and in Golubitsky and Guillemin's *Stable Mappings and their Singularities* [G5]. These works are quite difficult on account of their sophisticated mathematical terminology and notation; our approach will be more pedestrian, emphasising intuition rather than mathematical rigour.

We shall be concerned with the classification of critical points, for parameter and image spaces of low dimensionality, based on the use of non-singular analytic changes of the coordinates in image space and parameter space to reduce the description (5.1) to a simple *canonical form*. Two descriptions are regarded as *equivalent* in the neighbourhood of a critical point if they can both be reduced to the same canonical form.

Without loss of generality, the coordinate systems we begin with can be assumed to be chosen so that the point under investigation is the origin of the coordinate system in parameter space and its image is the origin of the coordinate system in image space. The Taylor series for the allowed coordinate changes can then be expressed by the Taylor series expansions of (4.2) and (4.3):

$$x'^i = A^i_j x^j + A^i_{jk} x^j x^k + \ldots \tag{5.2}$$

$$u'^\alpha = B^\alpha_\beta u^\beta + B^\alpha_{\beta\gamma} u^\beta u^\gamma + \ldots \tag{5.3}$$

in which the matrices appearing in the first terms are non-singular.

A critical point may be *stable* or *unstable*. It is defined to be unstable if the description can be made regular by a small *perturbation* $(f^i(u^\alpha) \to f^i(u^\alpha) + \epsilon^i(u^\alpha)$, the functions ϵ^i being 'infinitesimally small'). The significance of stability will become clear from the particular examples that follow.

5.1 Curves in Two Dimensions
Let

$$x^\alpha = f^\alpha(u) \quad (\alpha = 1, 2) \tag{5.4}$$

be a parametric description of a curve in a two-dimensional space. Given any point P on the curve, we can choose the coordinate system and the parametrisation so that P is the origin and corresponds to $u = 0$. If the curve is analytic at P, this means that a Taylor expansion about P is valid:

$$x^\alpha = k^\alpha_1 u + k^\alpha_2 u^2 + k^\alpha_3 u^3 + \ldots \tag{5.5}$$

The curve is singular at P (P is a critical point) if $k^\alpha_1 = 0$. Otherwise, it is regular at P.

For example, if the space is an affine plane (or a Euclidean plane) the coordinate system can be an affine system (§2.10), and (5.5) will then be written in the form

$$\mathbf{r} = \mathbf{k}_1 u + \mathbf{k}_2 u^2 + \ldots \tag{5.6}$$

The vector $\mathbf{k}_1$ is a *tangent* vector to the curve at P and the tangent line that touches the curve at P is given by the parametric equation $\mathbf{r} = \mathbf{k}_1 u$. In general, k_1 and k_2 will be linearly independent, and the affine coordinate system can be chosen so that they are unit vectors along the axes. Equation (5.6) then becomes $x = u\phi$, $y = u^2\psi$ where ϕ and ψ are analytic functions of u satisfying $\phi(0) = \psi(0) = 1$. Changing the parameter to $u' = u\phi$ we have $x = u'$, $y = u'^2 \chi(u')$ where χ is analytic with $\chi(0) = 1$. In terms of the new parameter,

$$\mathbf{r} = \begin{pmatrix} u \\ u^2 + 0\,(3) \end{pmatrix} \tag{5.7}$$

Thus, in the immediate vicinity of an ordinary regular point, any analytic curve resembles the curve $x^2 = y$. A special case arises if k_1 and k_2 are linearly dependent $(\mathbf{k}_2 = \alpha\mathbf{k}_1$ or $\mathbf{k}_2 = 0)$. If in this case k_1 and k_3 are linearly independent, the coordinate system can be chosen so that they are unit vectors along the axes. By changing the parameter in analogy with the previous example, we get

$$\begin{pmatrix} u \\ u^3 + 0\,(4) \end{pmatrix} \tag{5.8}$$

for the components of the position vector. This kind of regular point is an ordinary point of *inflexion*. In its immediate vicinity, the curve resembles the curve $x^3 = y$.

If P is a critical point, $k_1 = 0$. In general, k_2 and k_3 will be linearly independent; they can be taken to be unit vectors along the coordinate axes, so that (5.6) becomes $x = u^2\,\phi$, $y = u^3\,\psi$, with $\phi(0) = \psi(0) = 1$. Change the parameter to $u\phi^{\frac{1}{2}}$ and change the y-axis, and we obtain the expression

$$\begin{pmatrix} u^2 \\ u^3 + 0\,(4) \end{pmatrix} \tag{5.9}$$

for the position vector. The curve resembles $x^3 = y^2$ (Fig. 50). The point P is a simple first order *cusp*.

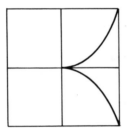

Figure 50

The generalization of the above arguments is now apparent. If $k_1 = k_2 = \ldots k_{n-1} = 0$, $k_n \neq 0$, $k_{n+1} \ldots k_{m-1}$ each proportional to k_n or zero, and $k_m \neq 0$, we can choose the affine coordinate system and the parametrisation so that the position vector is expressed in the form

$$\begin{pmatrix} u^n \\ u^m + 0\,(m+1) \end{pmatrix} (m > n) \tag{5.10}$$

For $n > 1$, we have a critical point of *order* $n - 1$ and *type* m. The case $n = 2$, $m = 4$ for example gives a 'ramphold' cusp (Greek ραμφωδης 'beak-shaped'), shown in Fig. 51.

In the above arguments, the image-space was taken to be a plane and its coordinate system was an affine coordinate system. The allowable coordinate changes were therefore restricted to linear transformations (first term in (5.3)). If we employ curvilinear coordinates (as we must if the image space is a surface

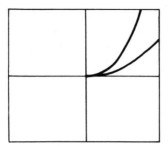

Figure 51

that is not a plane), the application of more general non-singular analytic changes of the coordinate system, in conjunction with parameter changes, enables us to remove most (and in some cases all) of the higher order terms $0\,(m+1)$ in (5.10). For example, for a regular point, (5.10) is $x = u$, $y = u^m \chi\,(\chi\,(0) = 1)$, and we can apply the transformation $y' = y - x^m \chi(x)$. We arrive at the *canonical form*

$$\begin{pmatrix} u \\ 0 \end{pmatrix} \tag{5.11}$$

Thus, all regular points are *equivalent* with respect to the classification scheme defined in the introduction to this chapter. The coordinate transformation can be regarded passively (as a change of curvilinear coordinate system) or *actively* as a mapping of a neighbourhood of P onto itself which specifies a deformation of the curve. Now consider the first order cusp of type m, for which (5.10) is $x = u^2$, $y = u^m \chi$. By subtracting a power series in x from y, we can eliminate all the even powers of u in y. Then $y = u^m v\,(u^2)$ where M is odd and v is an analytic function with $v\,(0) \neq 0$. A further coordinate change $y' = yv^{-1}\,(x)$ then gives the canonical form

$$\begin{pmatrix} u^2 \\ u^m \end{pmatrix} \quad \text{(M odd)} \tag{5.12}$$

for any first order cusp. In particular, any first order cusp of even type (such as the ramphoid cusp) in an affine plane can be deformed, by a non-singular analytic mapping applied to its neightbourhood, to a cusp of odd type. For second and higher order critical points, the canonical forms and the methods of obtaining them become more intricate.

Every critical point of a curve in a two-dimensional space is *unstable*. We shall demonstrate this for two simple cases. Consider the deformation of the simple first order cusp, that takes the canonical form to the form

$$\begin{pmatrix} u^2 \\ u^3 - eu \end{pmatrix} \tag{5.13}$$

If the constant ϵ is not zero, the curve has no critical point in the neighbourhood of P. For positive ϵ, the two values $u = \pm\sqrt{\epsilon}$ of the parameter correspond to the same point $(\epsilon, 0)$ on the curve — the curve has a double point or *node*. We see that a simple first order cusp can be 'perturbed' to produce a loop in the curve (Fig. 52).

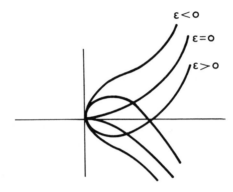

Figure 52

The canonical form for the critical points with $n = 3, m = 4$ is in general $x = u^3$, $y = u^4 + u^5$. In special cases it is $x = u^3, y = u^4$. For an arbitrary perturbation, the appeance of the resulting regular curve is shown in Fig. 53a. For particular choices of the perturbation, the critical point is revealed as a compound of two coincident simple first order cusps. This is the form, for instance, of the curve

$$\begin{pmatrix} x \\ y \end{pmatrix} = \begin{pmatrix} u^3 - 3\epsilon u \\ u^4 - 2\epsilon u^2 \end{pmatrix} \tag{5.14}$$

The situation illustrated in Fig. 53b will be encountered again in §5.9.

A more thorough mathematical treatment of the topics indicated in this section can be found in Coolidge [C6] or in Semple and Kneebone [S7].

 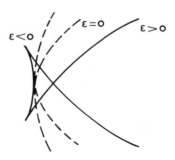

Figure 53

5.2 Space Curves

The nature of the critical points of curves in three-dimensional space can be investigated by extensions of the methods discussed in the previous section. We shall not do this; we simply consider the form of the ordinary regular point for a curve in affine or Euclidean 3-space, referred to an affine coordinate system. The parametric equation has the form (5.6), but now the vectors are regarded as vectors in 3-space. In general k_1, k_2 and k_3 will be linearly independent. The coordinate system can be chosen so that they are unit vectors along the axes. Then the parametric equation becomes $x = u\phi$, $y = u^2 \psi$, $z = u^3 \chi$, where ϕ, ψ and χ are analytic functions of the parameter satisfying $\phi(0) = \psi(0) = \chi(0) = 1$. Change the parameter to $u' = u\phi$, and we obtain (dropping the prime) $x = u$, $y = u^2 \eta$, $z = u^3 \zeta$ ($\eta(0) = \zeta(0) = 1$). The third order term in y can be eliminated by adding a multiple of z to y. We then obtain the form

$$\begin{pmatrix} u \\ u^2 + 0\,(4) \\ u^3 + 0\,(4) \end{pmatrix} \tag{5.15}$$

The curve resembles the curve $y = x^2$, $z = x^3$ in the vicinity of any regular point.

The parametric equation $\mathbf{r} = \mathbf{r}(s)$ of a curve in Euclidean 3-space can be referred to the coordinate system whose axes are along the vectors $\mathbf{t}$, $\mathbf{n}$ and $\mathbf{b}$. By applying the Serret-Frenet formulae to the Taylor expansion of $\mathbf{r}(s)$ we find

$$\begin{pmatrix} x \\ y \\ z \end{pmatrix} = \begin{pmatrix} s + \ldots \\ \dfrac{\kappa}{2} s^2 + \ldots \\ \dfrac{\kappa \tau}{6} s^3 + \ldots \end{pmatrix} \tag{5.16}$$

Therefore, at an ordinary regular point of a curve in Euclidean space, the projection of the curve on its osculating plane (the xy-plane) has an ordinary regular point, the projection on its rectifying plane (the xz-plane) has an inflexion, and the projection on its normal plane has a simple first order cusp.

5.3 Scalar Field in Two Dimensions

The Taylor expansion about a critical point of a mapping $x = f(u, v)$ from two dimensions to one has the form

$$x = f_2 + f_3 + f_4 + \ldots \tag{5.17}$$

where $f_n = f_n(u, v)$ is a homogeneous polynomial of degree n in the parameters u and v. For example

$$f_2 = Au^2 + 2Buv + Cv^2 \tag{5.18}$$

If this term is not identically zero, the critical point can be classified according to the signature of the matrix $\begin{pmatrix} AB \\ BC \end{pmatrix}$.

(i) *Both eigenvalues positive.* A (linear) parameter change can convert the matrix to the unit matrix. Then $x = u^2 \phi + v^2 \psi$, $\phi(0) = \psi(0) = 1$. The parameters can be changed to $u\phi^{\frac{1}{2}}$ and $v\psi^{\frac{1}{2}}$. We obtain the canonical form

$$x = u^2 + v^2 \tag{5.19}$$

The critical point is a *minimum* of the scalar field x. Its form is illustrated, as a graph of x plotted against u and v, and by means of contours, in Fig. 54

 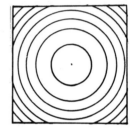

Figure 54 $x = u^2 + v^2$

(ii) *Both eigenvalues negative.* The critical point is a minimum of the scalar field.

(iii) *Eigenvalues opposite in sign.* The canonical form is

$$x = u^2 - v^2, \tag{5.20}$$

corresponding to a *saddle point* of the scalar field, illustrated in Fig. 55 (the dotted contours correspond to negative values of x). An alternative canonical form for the saddle-point is

$$x = uv \tag{5.21}$$

obtained from (5.20) simply by the transformation $u \rightarrow u + v$, $v \rightarrow u - v$ (Fig. 56).

The saddle point, the maximum, and the minimum are the only kinds of *stable* critical point that can occur in a scalar field in two dimensions.

(iv) *One eigenvalue zero.* If one eigenvalue is zero, the third order term cannot be completely eliminated. In the general case, with one eigenvalue zero, we can arrive at the canonical form

$$x = u^2 + v^3 \tag{5.22}$$

The form of the field in the neighbourhood of such a critical point is indicated in Fig. 57. Its instability is demonstrated by considering the forms of the fields $x = u^2 + v^3 \pm \epsilon v$ (ϵ positive). With the positive sign, there is no

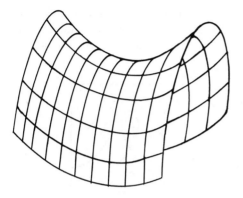
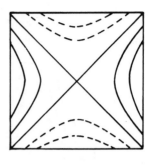

Figure 55 $x = u^2 - v^2$

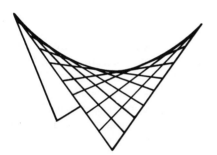
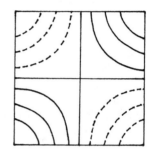

Figure 56 $x = uv$

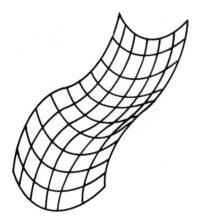

Figure 57 $x = u^2 + v^3$

critical point. With the negative sign, there is a minimum at $(0, \sqrt{\epsilon})$ and a saddle-point at $(0, -\sqrt{\epsilon})$ (Fig. 58). Thus a singularity with the canonical form (5.22) can be regarded as a degenerate critical point where a minimum and a saddle-point coalesce.

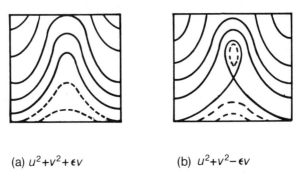

(a) $u^2 + v^2 + \epsilon v$ (b) $u^2 + v^2 - \epsilon v$

Figure 58

Special cases can arise, which prevent a critical point with one zero eigenvalue of $a_{\alpha\beta}$ from being reduced to the canonical form (5.22). In fact, $u^2 + v^3$ belongs to a set of canonical forms $u^2 \pm y^{2m}$ and $u^2 + v^{2m+1}$. Their general appearance is indicated in Fig. 59. A critical point, of course, corresponds to a root of the simultaneous equations $\partial f/\partial u = \partial f/\partial v = 0$. The index m is related to the *multiplicity* of this root. The degeneracy is removed by a perturbation. For example, consider $x = u^2 + v^4 + \epsilon(u, v)$. The system of equations $x_u = 2u + \epsilon_u = 0$, $x_v = 4v^3 + \epsilon_v = 0$ has three roots in the neighbourhood of the origin. If one is real and two form a conjugate complex pair, this means that the perturbation has converted the singularity of $u^2 + v^4$ to an ordinary minimum. If we have three real roots, it has converted the singularity to two minima and a saddle-point. The new forms of the singularities of Fig. 59, when subjected to perturbations so that the roots of $x_u = x_v = 0$ are all real, are shown in Fig. 60.

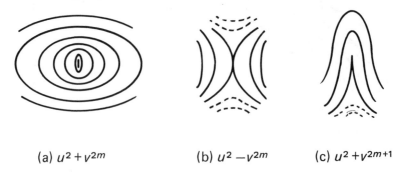

(a) $u^2 + v^{2m}$ (b) $u^2 - v^{2m}$ (c) $u^2 + v^{2m+1}$

Figure 59

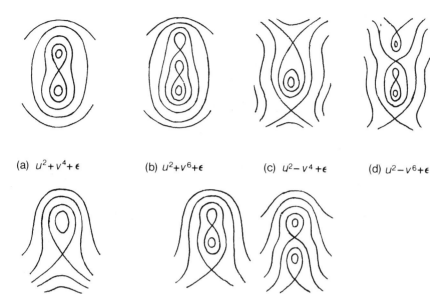

(a) $u^2+v^4+\epsilon$ (b) $u^2+v^6+\epsilon$ (c) $u^2-v^4+\epsilon$ (d) $u^2-v^6+\epsilon$

(e) $u^2+v^3+\epsilon$ (f) $u^2+v^5+\epsilon$ (two forms)

Figure 60

(v) *Second order terms absent.* If the quadratic terms in (5.17) are zero, we have to look to the cubic terms for the classification. We consider only the case where the cubic has distinct roots. They determine the directions $u : v$ of the zero contour through P. For three real roots, we obtain, in general, the canonical form $u(u^2 - v^2)$, or equivalently $u(u^2 - 3v^2)$. This is the 'monkey-saddle' (Fig. 61). Every vertical section through P has an inflexion at P.

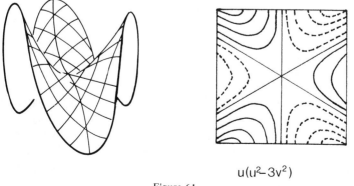

$u(u^2-3v^2)$

Figure 61

If only one root is real, we get the canonical form $u(u^2 + v^2)$ (Fig. 62).

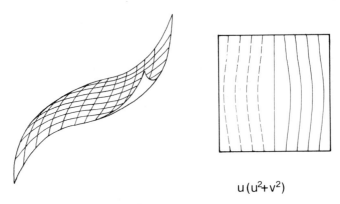

$$u(u^2 + v^2)$$

Figure 62

5.4 Mappings from Two Dimensions to Two Dimensions

We shall not any longer give detailed derivations. For more details, the reader is referred to Golubitsky and Guillemin [G5]. There are only two kinds of *stable* critical point for $M = N = 2$; the *fold point* (rank 1), with canonical form

$$\begin{pmatrix} x \\ y \end{pmatrix} = \begin{pmatrix} u \\ v^2 \end{pmatrix} \tag{5.23}$$

(Fig. 63a) and the *cusp on a fold line* (rank 0), with canonical form

$$\begin{pmatrix} x \\ y \end{pmatrix} = \begin{pmatrix} u \\ v^3 + uv \end{pmatrix} \tag{5.24}$$

(Fig. 64).

Fig 63b shows a perturbation in the neighbourhood of a fold point. The fold is not removed — it is *stable*.

A particular instance of a mapping with $M = N = 2$ is the orthographic projection of a surface on a plane. A *fold* is the edge of the shadow. Consider, for example, the projection of the surface $x = u$, $y = v^2$, $z = v$ on the (x, y)-plane (Fig. 65), and the similar projection of the surface $x = u$, $y = uv + v^3$, $z = v$ (Fig. 66). This is a *ruled surface*, the curves $v = $ const. being lines parallel to the (x, y)-plane. The planes $x = $ const. intersect the surface along the coordinate curves $u = $ const., which are cubics.

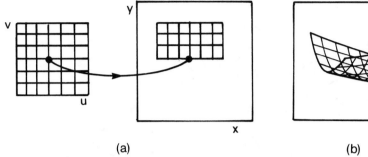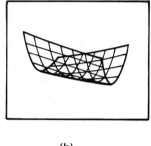

(a)   (b)

Figure 63

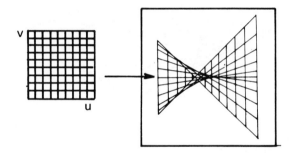

Figure 64

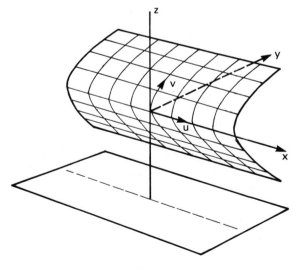

Figure 65

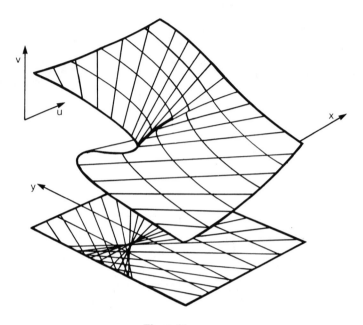

Figure 66

5.5 Surfaces in 3-space

Let us consider, first, surfaces in affine or Euclidean 3-space, referred to an affine coordinate system. The Taylor expansion of the parametric equation $\mathbf{r} = \mathbf{f}(u, v)$ about a point P on the surface (taken to be the origin of the affine system and of the intrinsic coordinate system u, v) is

$$\mathbf{r} = \mathbf{f}_1 + \mathbf{f}_2 + \mathbf{f}_3 + \dots \qquad (5.25)$$

where $\mathbf{f}_n = \mathbf{f}_n(u, v)$ is a homogeneous polynomial of degree n in u and v. For example,

$$\mathbf{f}_1 = \mathbf{k}_1 u + \mathbf{k}_2 v \qquad (5.26)$$

If P is a regular point, the two vectors $\mathbf{k}_1$ and $\mathbf{k}_2$ are linearly independent. The *tangent plane* at P is the plane through these two vectors, i.e. the plane given parametrically by the equation $\mathbf{r} = \mathbf{f}_1$. The affine coordinate system can be chosen so that these two vectors are unit vectors along the x and y axes. Then (5.25) becomes $x = u\phi$, $y = v\psi$, $z = \chi$ where ϕ, ψ and χ are analytic functions of u and v, $\phi(0, 0) = \psi(0, 0) = 1$, and χ has no constant or linear terms. Changing the parameters to $u\phi$ and $v\psi$, we obtain the expression

$$\begin{pmatrix} u \\ v \\ f(u, v) \end{pmatrix} \qquad (5.27)$$

for the position vector, in which $f(u, v)$ is an analytic function of u and v without constant or linear terms. The surface can now be expressed in Monge's form $z = f(x, y)$, in the neighbourhood of P. The xy-plane in this description is the tangent plane at P. The expansion of $f(x, y)$ begins with quadratic terms, so the various kinds of regular point of a surface correspond to the various kinds of critical point of a scalar field in two dimensions ($\S 5.3$. The variables x, u and v of that section are the z, x and y, respectively, of the present description). The important cases are as follows:

(i) $f(x, y)$ has a maximum or minimum at the origin (Fig. 54). The surface is *synclastic* at P.

(ii) $f(x, y)$ has a saddle point at the origin (Fig. 55). The surface is *anticlastic* at P.

(iii) $f(x, y)$ has a critical point with canonical form $x^2 + y^3$ (Fig. 57). The point P is a *spinodal point* (or 'parabolic point') of the surface.

The spinodal points lie on a spinodal curve, which separates the synclastic and anticlastic regions. In $\S 2.7$, the concept of synclastic and anticlastic regions of a surface was defined in terms of the *Euclidean* idea of curvature. We now see that it is a more primitive, *affine* concept.

The canonical forms for critical points of a surface are obtained by applying non-singular analytic transformations to the coordinates as well as the parameters. The actual processes of reduction to canonical form are quite intricate, and will not be given here. We shall simply illustrate the results.

A critical point of rank 1 occurs when the two vectors $\mathbf{k}_1$ and $\mathbf{k}_2$ of (5.26) are linearly dependent. In general, the canonical form

$$\begin{pmatrix} x \\ y \\ z \end{pmatrix} = \begin{pmatrix} u \\ v^2 \\ uv \end{pmatrix} \tag{5.28}$$

can be obtained. The non-parametric description is $z^2 = x^2 y$. The form of the surface is illustrated in Fig. 67. This kind of critical point is called a *pinch point*. The surface has a line of self-intersection that terminates at the pinch point. Note that the embedding of a projective plane in 3-space illustrated in Fig. 15 gives rise to a surface with two pinch' points.

A pinch point is stable.

An interesting aspect of Fig. 67 is revealed if we consider the intersections with a family of parallel planes, for example the planes parallel to the plane $x = y$ (Fig. 68). We get a family of curves displaying the transition from a plane curve with a loop to one without a loop, via a series of perturbations of a cusp. Compare this with Fig. 52.

In addition to the pinch point, which is the general case, we can have critical points of rank 1 which cannot be reduced to the canonical form (5.28). Two such special cases are the *conic node* with canonical form

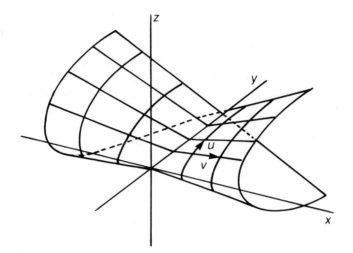

Figure 67

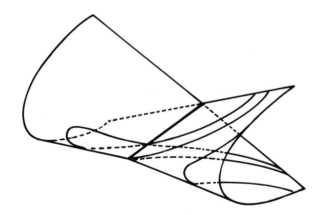

Figure 68

$$\begin{pmatrix} u \\ uv^2 \\ uv \end{pmatrix}$$

(5.29)

(i.e. $z^2 = xy$) (Fig. 69) and the *cuspidal edge* with the canonical form

$$\begin{pmatrix} u \\ v^2 \\ v^3 \end{pmatrix}$$

(5.30)

shown in Fig. 70.

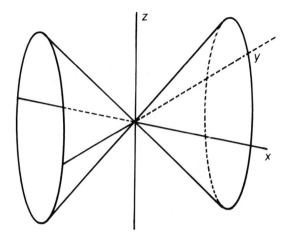

Figure 69

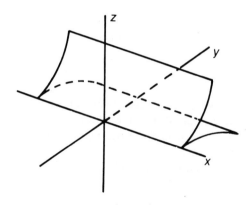

Figure 70

The cuspidal edge occurs when a surface is generated by the tangent lines to an analytic curve in Euclidean or affine 3-space (Fig. 71). We shall give a non-rigorous demonstration of this based on the form (5.15) for the curve, ignoring the fourth-order terms. A tangent vector at the point u has the components $(1, 2u, 3u^2)$ (obtained by differentiation of $\mathbf{r}$ with respect to the parameter). The surface generated by the tangent lines is therefore described by the parametric form $\mathbf{r} = (u + v, u^2 + 2uv, u^3 + 3u^2v)$. Change the parameter u to $u' = u + v$, and then make the coordinate change $y' = x^2 - y, z' = \frac{1}{2}(z + 2x^3 - 3y)$. We obtain the canonical form (5.30) for the cuspidal edge.

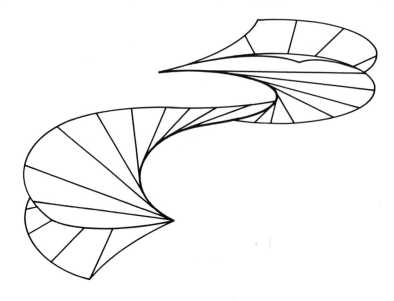

Figure 71

5.6 Critical Points of Algebraic Curves and Surfaces

An *algebraic* curve in an affine or Euclidean plane is a curve described by an equation $\phi(x, y) = 0$, where ϕ is a polynomial . The *degree* of the curve is the degree of the polynomial. For degrees, 1, 2, 3 etc. we speak of lines, conics, cubic curves, etc. Choosing the origin at a point P on the curve, the equation takes the form

$$\phi = f_1 + f_2 + \ldots f_n = 0 \qquad (5.31)$$

where f_i is a *homogeneous* polynomical of degree i in x and y (that is, f_1 is a linear combination of x and y, f_2 is a linear combination of x^2, xy and y^2, and so on). The curve is *regular* at P if f_1 is not identically zero. Otherwise (i.e. if this term is absent from the equation), the point P is a singular point of the curve.

Let Q be an arbitrarily chosen point in the plane, other than P. If its co-ordinates are (x, y), the coordinates of all the points on the line PQ have the form $(\lambda x, \lambda y)$. Therefore, the intersections of the curve with the line PQ are given by the roots of the equation

$$\lambda f_1(Q) + \lambda^2 f_2(Q) + \ldots \lambda^n f_n(Q) = 0 \qquad (5.32)$$

The line cuts the curve in two coincident points at P if the equation has a double root at $\lambda = 0$. This occurs when $f_1(Q) = 0$. Therefore, if P is a regular point, the equation of the *tangent* at P is $f_1 = 0$ (Fig. 72a).

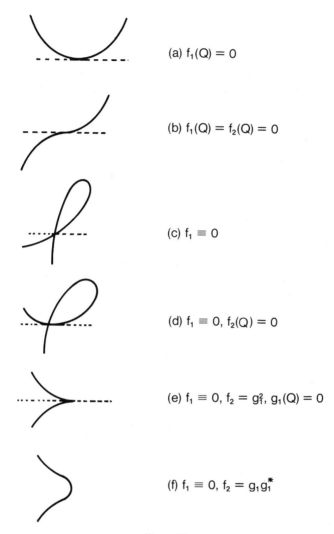

(a) $f_1(Q) = 0$

(b) $f_1(Q) = f_2(Q) = 0$

(c) $f_1 \equiv 0$

(d) $f_1 \equiv 0,\ f_2(Q) = 0$

(e) $f_1 \equiv 0,\ f_2 = g_1^2,\ g_1(Q) = 0$

(f) $f_1 \equiv 0,\ f_2 = g_1 g_1^{*}$

Figure 72

If $f_1(Q) = f_2(Q) = 0$, the tangent touches the curve at *three* coincident points. P is an *inflexion* of the curve (Fig. 72b). The obvious generalisation to tangents with higher order contact at P, is immediately apparent.

If P is singular point, the first term is absent from (5.32). Then *every* line through P cuts the curve in (at least) two coincident points at P (Fig. 72c). The curve has a double point, or *node* at P. If, in this case, Q satisfies $f_2(Q) = 0$ we will have three coincident points at P. PQ is in this case tangential to one of the branches of the curve that pass through P (Fig. 72d). Therefore, the equation

$f_2 = 0$ is the equation of the two tangent lines at the node. If the linear factors of f_2 are identical $(f_2 = g_1^2)$ we have only one tangent at P, and we have a *cusp* (Fig. 72e). If the linear factors of f_2 are conjugate complex $(f_2 = g_1 g_1{}^*)$ there is no real tangent at P, and in fact P is isolated from the rest of the curve (Fig. 72f). The last three types of curve have been encountered already in the context of parametric descriptions (Fig. 52). The three curves are the cubic curves

$$z^2 = x^2 (x - \epsilon)$$

with ϵ positive, zero or negative. In the latter case, the point $(0, \epsilon)$ is isolated from the rest of the curve.

Proceeding in the same way, the various forms that arise when both f_1 and f_2 are identically zero can be investigated. We shall leave the topic at this point and proceed to investigate algebraic *surfaces*.

An algebraic surface in affine or Euclidean 3-space is described by a polynomial equation $\phi(x, y, z) = 0$. For degrees 1, 2, 3, etc. we speak of planes, quadrics, cubic surfaces, etc. Let the origin be a point P on the surface, so that the equation has the form (5.31), where each term is a homogeneous polynomial in the *three* coordinates. The intersections of the surface with a line PQ are given by the roots of an equation of the form (5.32). If P is a regular point of the surface, PQ will touch the surface (i.e. cut it at two coincident points) at P if $f_1(Q) = 0$. It follows that $f_1 = 0$ is the equation of the *tangent plane* at P. The surface intersects the tangent plane in a plane curve with a singularity at P. The surfaces is anticlastic, spinodal or synclastic at P according to whether this singularity is a double point, a cusp, or an isolated point.

If the surface is singular at P, the first term in (5.32) is absent; every line through P touches the curve in two coincident points. Those with 'tritactic' contact at P are given by the condition $f_2(Q) = 0$. The equation $f_2 = 0$ is an equation of a quadric *cone*, the *tangent cone* to the surface at P. The surface has a *conic node* at P. An example of such a point is obtained by considering the surface generated by rotating the curve labelled $\epsilon > 0$ in Fig. 52 about the x-axis.

A tangent cone at a singular point can be degenerate, consisting of a *pair of planes* $(f_2 = g_1 h_1)$, which may be real and distinct or conjugate complex, in which case P is called a *binode*, or they may coincide $(f_2 = g_1^2)$ in which case P is called a *unode*. An example of a binode with conjugate complex tangent planes is the origin on the surface $z^3 = x^2 + y^2$ (the imaginary tangent planes are $x = \pm iy$) generated by rotating the curve in Fig. 50 about its tangent vector. An example of a binode with two real tangent planes is given by the surface $z^3 = xy$ (Fig. 73).

Examples of unodes are: a point on a *cuspidal edge* (Figs. 70, 71) and a *pinch point* (Fig. 67). Note that the y-axis is a line of binodes of the algebraic surface $z^2 = x^2 y$ of this figure. On the positive half of the axis they have a pair of real

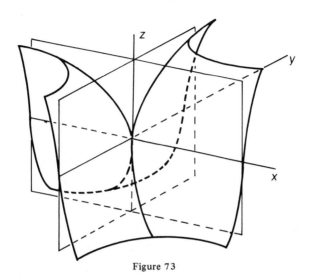

Figure 73

tangent planes, and on the negative half they are isolated from the rest of the surface, and have conjugate complex tangent planes.

Parametric descriptions exist for any plane algebraic curve. More precisely, for any chosen point P on a plane algebraic curve, the 'branch' of the curve on which P lies can be described by Taylor series expansion of the coordinates [C6]. Similarly, parametric descriptions exist for any algebraic surface. These theorems explain why the phenomena that we encountered when studying analytic curves and surfaces appeared again, from a different viewpoint, in the study of algebraic curves and surfaces. The contrast between the two approaches is epitomised by the nature of a simple double point of a plane curve, which is a singularity from the algebraic viewpoint, but not from the viewpoint of parametric description.

5.7 Envelopes

A mapping with M = N = 2 can describe a family of curves in a two-dimensional space:

$$\left.\begin{array}{c} x = f(u,v) \\ y = g(u,v) \end{array}\right\}$$ (5.33)

For each fixed value of v, we have a parametric description (with u as the parameter) of one curve of the family. The *envelope* of a family of curves described by (5.33) is the *fold line* of this mapping (if it exists), given by the equation

$$f_u g_v - f_v g_u = 0$$ (5.34)

The allowed parameter changes are restricted to the form

$$\left.\begin{aligned} u' &= b\,(u,\,v) \\ v' &= c\,(v) \end{aligned}\right\} \tag{5.35}$$

since a more general change in v would give curves v = const. not belonging to the family.

Let the origins of parameter space and image space correspond to a point on the envelope. The origin is then a fold-point of the mapping (5.33) so we have, according to (5.23), if the origin is not a critical point of the envelope,

$$\left.\begin{aligned} x &= \xi \\ y &= \eta^2 \end{aligned}\right\} \tag{5.36}$$

where ξ and η are some functions of u and v. The envelope is then the x-axis. The parameter u can be changed, according to (5.35), so that $u = \xi$. We then have $x = u$, $y = \eta^2$, $\eta = (\alpha u + \beta v)\,\phi(u,\,v)$. We assume a general case, so that α and β are not zero, and ϕ has a non-vanishing constant term, which can be taken to be 1. The transformations $x' = x/\alpha$, $u' = \alpha u$, $v' = \beta v$ then lead to the form $x = u$, $y = (u + v)\,\zeta\,(u,\,u + v)$ (where the primes have been omitted). Changing the y-coordinate to y' given by $y = y'\zeta\,(x,\,y')$, we finally obtain the canonical form

$$\left.\begin{aligned} x &= u \\ y &= (u + v)^2 \end{aligned}\right\} \tag{5.37}$$

for a family of curves in the neighbourhood of a general point on the envelope (Fig. 74).

Figure 74

An alternative canonical form is obtained by applying the further transformation $y' = y - x^2$. We get

$$\left.\begin{aligned} x &= u \\ y &= uv + v^2 \end{aligned}\right\} \tag{5.38}$$

The Jacobian (5.34) for this canonical form is $u + 2v$, so the envelope is $y = -x^2/4$ (Fig. 75).

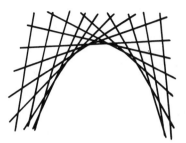

Figure 75

The envelope of a family of curves described by (5.33), in a Euclidean plane, can be visualised as the edge of the shadow of the orthographic projection of the surface

$$\left.\begin{array}{l} x = f(u,v) \\ y = y(u,v) \\ z = v \end{array}\right\} \qquad (5.39)$$

in Euclidean 3-space, on the (x, y)-plane. It follows that, if the family of curves (5.33) is described in the alternative form

$$\phi(x, y, v) = 0 \qquad (5.40)$$

(corresponding to the description $\phi(x, y, z) = 0$ of the surface (5.39)), then the envelope is given by the equation

$$\partial\phi/\partial v = 0 \qquad (5.41)$$

There are various ways in which a cusp can arise on an envelope. Fig. 76a is obtained by the orthographic projection of the ruled cubic surface shown in Fig. 66, the family of curves being images of the surface rulings. Fig. 76b corresponds to a projection of the cubic sections of the surface, on a differently oriented plane. Two cusps on an envelope are evident at A and B in Fig. 77. They are of the same canonical types as Fig. 76a and Fig. 76b respectively. It is not difficult to see how the family of curves in this figure can be regarded as a projection on the plane of a family of curves on a surface (like streamlines on the surface of a whirlpool).

Fig. 78 shows a typical example of the appearance of a family of curves when a cusp on the envelope coincides with a cusp on one of the curves of the family. The figure can be regarded as arising from the projection of a family of curves on a surface with a 'cross-cap' (c.f. Fig. 67). Some very elegant computer-generated figures of this type — that is, projections onto a plane of families of curves on surfaces with singularities — are to be found in Woodcock and Poston [W10].

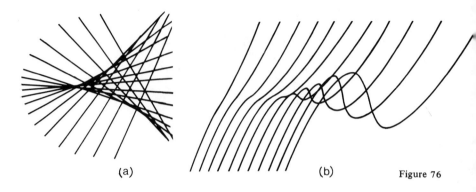

(a)　　　　　(b)　　　Figure 76

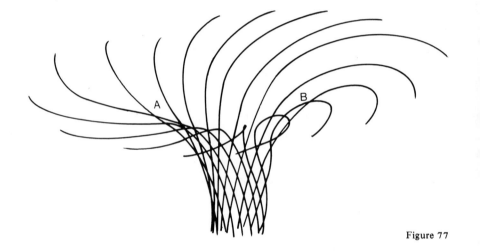

Figure 77

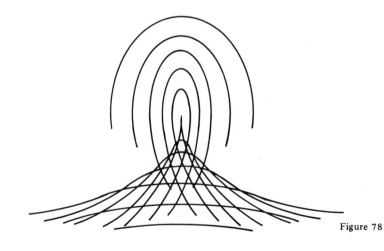

Figure 78

5.8 Removal of Critical Points and Multiple Points

Consider a cusp on a plane curve. The canonical form is $x = u^2, y = u^3$. Defining $z = u$, we obtained a *regular* curve in 3-space, whose orthographic image on the (x, y)-plane is the original plane curve with a cusp (Fig. 79). Thus we see the

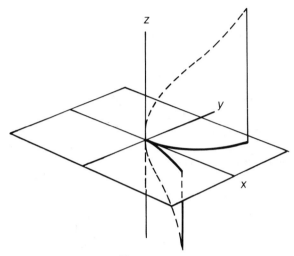

Figure 79

possibility of removal of critical points by extending the dimensionality of the image space. Multiple points (nodes) can be removed in the same way. For example, the plane curve $x = u^2, y = u^3 - u$ has a double point at $u = \pm 1$. Defining $z = u$, we obtain a curve in 3-space without a double point (Fig. 80).

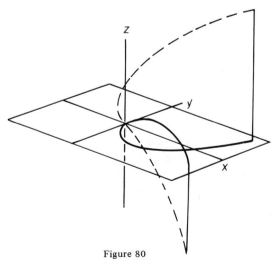

Figure 80

The generalisation of the procedure is straightforward. A mapping from M dimensions to N with a critical point of rank r at the origin, has a canonical form

$$\left. \begin{array}{l} x^i = u^i \, (i = 1 \ldots r) \\ x^j = \phi^j(u) \, (j = r + 1, \ldots N) \end{array} \right\} \tag{5.42}$$

where the functions ϕ^j have no linear terms. We extend the dimensionality of the image space to $M + N - r$, according to

$$x^k = u^{k - N + r} \, (k = N + 1, \ldots M + N - r) \tag{5.43}$$

For example, consider a fold in a mapping with $M = N = 2$. The canonical form is (5.23), $x = u$, $y = v^2$. Then, defining $z = v$ gives a *regular* surface in 3-space, whose orthographic projection gives the image with a fold (Fig. 65). Similarly, a fold line with a cusp, $x = u$, $y = uv + v^3$, is removed by defining $z = v$. The situation is that of Fig. 66.

The removal of a 'cross-cap' on a surface in space, and its associated line of double points, is removed if we extend the dimensionality of the space to 4, to obtain the regular surface

$$\left. \begin{array}{l} x = u \\ y = v^2 \\ z = uv \\ w = v \end{array} \right\} \tag{5.44}$$

As a simple example with $M > N$, consider the simple minimum of a scalar field (5.19) $x = u^2 + v^2$. This critical point is removed by introducing *two* extra dimensions,

$$\left. \begin{array}{l} x = u^2 + v^2 \\ y = u \\ z = v \end{array} \right\} \tag{5.45}$$

This is, of course, the regular surface in 3-space that represents the scalar field (Fig. 54).

5.9 Blum's Medial Axis Description

Consider a two-dimensional wave-motion, for waves propagating with uniform velocity. The form of a wave-front at a particular time determines a set of *rays*, which are the straight lines orthogonal to the wavefront. The successive wavefronts at later (or earlier) times are the curves that intersect the rays orthogonally. The envelope of the rays is the *caustic curve* associated with the wave pattern (indicated by the dotted line in Fig. 81). It is the evolute (locus of centres of curvature) for every wavefront.

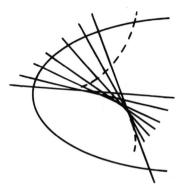

Figure 81

These geometrical properties of wave propagation are closely related to Blum's method of describing the forms of closed curves in a Euclidean plane [B3, B4]. The outline of any object in the plane is considered to be a wavefront at some time $t = 0$, and the wave is imagined to propagate into the interior of the object with uniform velocity. If the outline is a smooth curve, the wavefront eventually develops second order cusps at the centres of curvature for which the curvature is a maximum, and at a later time has a pair of cusps and a double point as in Fig. 53b. This is shown in Fig. 82, with the caustic curve indicated by a dotted line. The locus of double points on the wavefronts is

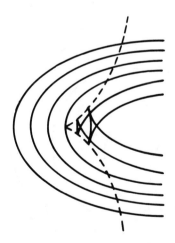

Figure 82

Blum's *medial axis* of the original outline (Fig. 83). The parametric expression

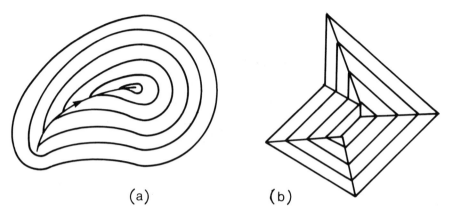

(a)   (b)

Figure 83

$x = \xi(t), y = \eta(t)$ of the medial axis (with time as the parameter) *contains a complete description* of the form of the original outline. The functions $\xi(t)$ and $\eta(t)$ will not in general be single-valued — medial axes can have a branched structure, as in Fig. 83b.

The regeneration of an outline from the medial axis description can be obtained by imagining moving sources traversing the medial axis in a direction opposite to that of the process that gave rise to the axis. That is, if $x = \xi(t)$, $y = \eta(t)$ is regarded as the position of a source at time $-t$, the wavefront produced at time $t = 0$ will coincide with the original outline, which now has a description as the envelope of a family of circles (Fig. 84)

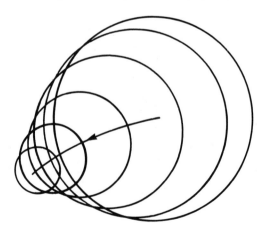

Figure 84

Blum has put forward a theory of visual perception based on the medial axis concept. The medial axis description was employed by Hilditch [H9] in computer analysis of chromosome photographs.

5.10 Thom's Catastrophe Theory

René Thom's theory of morphogenesis is a remarkable example of the application of the theory of singularities of mappings. It is an attempt to elucidate the general dynamical principles underlying the changes of form that take place in nature. The emphasis is on embryology, but the methods are applicable to the evolution of galaxies, the breaking of ocean waves, the formation of bubbles, geological structures − in fact, to any dynamical process where the characteristics of a form undergo abrupt changes. The definitive exposition of the theory is Thom's *Structural Stability and Morphogenesis* [T2].

The fundamental idea exploited in the theory is that of a catastrophe. Suppose the stable configuration of a physical system is determined by a minimum of a potential function, and suppose that the form of the potential function undergoes a gradual development in time. Stability is lost when the minimum ceases to exist, and we then have a 'catastrophe': the system undergoes a rapid sequence of changes until a new stable configuration is achieved. This can be illustrated by the following simple example:

Consider a particle in a one-dimensional space (coordinate χ) under the influence of the potential $V = \frac{1}{3}\chi^3 - a\chi$. As long as a is positive, the particle has a stable position at $\chi = \sqrt{a}$. If a changes, this stability ceases when $a = 0$, and a catastrophe occurs. If V has a more complicated form, resembling this one close to the origin, there may be other minima, and the particle will move rapidly until it finds one of them.

Multi-dimensional analogues of this situation are common in nature, when a form goes through an orderly sequence of changes under the influence of changes in the parameters that govern the physics that determine the form, until some unstable situation is reached. There is then a rapid transition to a new kind of form.

The simple example shown in Fig. 85 is taken from Sir James Jeans' *Astronomy and Cosmogony* [J2]. It represents the sequence of configurations of a rotating liquid mass (the computations were carried out by Jeans for the analogous two-dimensional case). For small angular velocities, the form is an ellipsoid, but as the velocity increases, a *catastrophe point* is reached. The form then rapidly passes through the sequence shown in the figure until a new stable configuration, consisting of a pair of smaller ellipsoids, is arrived at.

Returning now to the one-dimensional example, the potential $V = \frac{1}{3}\chi^3 - a\chi$ can be represented as a surface in a three-dimensional space with coordinates (χ, a, V). The projection of the minima and maxima of V on the (χ, a)-plane gives a curve, called the *catastrophe manifold* for the potential. It is the curve $\chi^2 = a$, with parametric description $\chi = u$, $a = u^2$. The mapping of a catastrophe

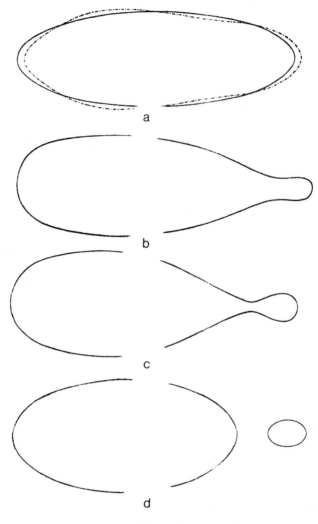

a

b

c

d

Figure 85

manifold on the one-dimensional a-space is the *catastrophe map*, which has a critical point at the origin (Fig. 86).

Now generalise this picture. Consider a dynamical system with N degrees of freedom, described by N generalised coordinates χ^i. The N-dimensional space is the *behaviour space* for the system. Suppose the behaviour of the system is governed by a potential $V(\chi^i, a^\alpha)$ which can change due to changes in a set of M parameters a^α. The M-dimensional space of these parameters is the *control space*. The stable configurations are determined by the condition

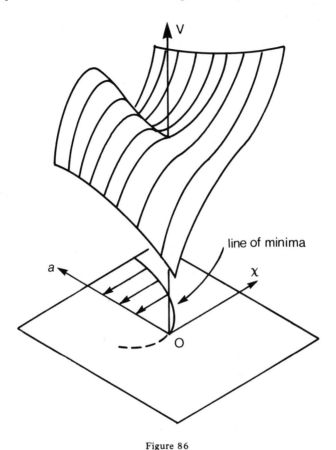

Figure 86

$\partial V/\partial \chi^i = 0$. They determine the points of an M-dimensional space immersed in the M + N dimensional space coordinatised by (χ^i, a^α). This M-space is the *catastrophe manifold*. The catastrophe map is the projection of this manifold on the control space. A *catastrophe* is a critical point of this map. An evolution of the system is brought about by gradual changes of the parameters a^α, represented by a curve in control space. When this curve reaches a catastrophe point, the stability of the system is lost, and an abrupt change to a new configuration will set in.

The study and classification of catastrophe situations is based on the classification of the types of critical points of mappings. The one-dimensional example and that taken from Jeans are examples of the simplest kind of catastrophe, the *fold* catastrophe. The fold catastrophe is the only kind that exists for a one-dimensional control space.

Consider now a one-dimensional behaviour space, in which the dynamics is governed by a *two*-dimensional control space. We take as an example the potential

$$V = \tfrac{1}{4}\chi^4 + \tfrac{1}{2}a\chi^2 + bx \qquad (5.46)$$

The catastrophe manifold is a surface in 3-dimensional (a, b, χ)-space, whose equation is

$$\frac{\partial V}{\partial \chi} = \chi^3 + a\chi + b = 0 \qquad (5.47)$$

The catastrophe map is the orthographic projection of this surface on the (a, b)-plane. The map is singular along the curve $(a/3)^3 + (b/4)^2 = 0$ which has a cusp at the origin (Fig. 87). The general points of this curve are catastrophes of the fold type. The cusp is a new type, called a Riemann-Hugoniot catastrophe. Take the point P in the figure as the starting-point of a dynamical process. The corresponding point vertically above P on the catastrophe manifold determines

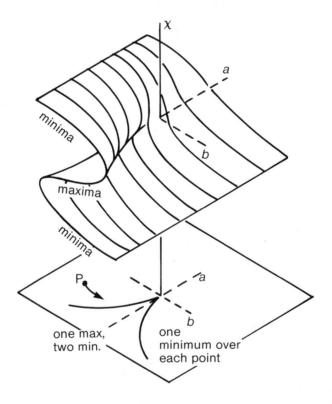

Figure 87

a single stable configuration χ (a minimum of the potential). As P moves to the right, parallel to the b-axis, the first branch of the cusped curve has no effect, but at the second branch the stability is lost and the value of χ must suddenly change (as we 'drop over the edge' on the catastrophe manifold). A different situation occurs as we move along the a-axis from positive a to the origin. We reach a *bifurcation* point where the future evolution of the system has two possible modes of development.

In higher dimensions the situations become increasingly difficult to visualise. Woodcock and Poston [W10] have provided a very powerful aid to an intuitive understanding of the geometry involved, in the form of a very beautiful collection of computer-generated pictures of two-dimensional catastrophe manifolds and sequences of cross-sections of higher-dimensional catastrophe manifolds. For a non-technical discussion of some of the applications of Thom's theory, the reader is referred to the article by Stewart [S17]. Many of the applications of the theory, particularly in biological morphogenesis, are due to Zeeman [Z1].

Catastrophe theory is actually a generalisation of the theory of envelopes, as is easily seen by noting that the envelope of a family of curves $\phi(x, y, u) = 0$ is the catastrophe set associated with the potentials $V(x, y, u) = \int \phi(x, y, u) du$, where u refers to a one-dimensional behaviour space and (x, y) refers to a two-dimensional control space.

5.11 Vector Field in Two Dimensions

A vector field in a two-dimensional space can be described by a mapping with $M = N = 2$. For consistency with the usual notation for fluid flow, we shall denote the coordinates of parameter space by (x, y) and those of the image space (i.e. the vector components) by (u, v). Note that this is the reverse of the notation we have previously employed.

A vector field in two dimensions is then specified by

$$\left.\begin{aligned} u &= f(x, y) \\ v &= g(x, y) \end{aligned}\right\} \tag{5.48}$$

We assume the functions f and g are analytic. The concept of singularity for a vector field differs from the concept of singularity that we discussed hitherto: a point in a vector field is *singular* if the vector at the point is zero.

Denoting the vector whose components are (u, v) by $\mathbf{u}$, the Taylor expansion of (5.48) at a singular point P, taken to be the origin of the x and y coordinates, has the form

$$\mathbf{u} = \mathbf{u}_1 + \mathbf{u}_2 + \ldots \tag{5.49}$$

where the components of the vector $\mathbf{u}_n$ are homogeneous polynomials of degree n in x and y. Thus

$$\mathbf{u}_1 = \mathbf{M}\mathbf{r} \tag{5.50}$$

where **r** is the position vector and M a matrix independent of **r**.

We shall discuss only stable first-order critical points. They are those for which the matrix M appearing in (5.50) is non-singular. We shall classify the coordinates according to the canonical forms of this matrix. The matrix can be reduced to canonical form by an affine transformation of the coordinates ($\mathbf{r}' = A\mathbf{r}$, $\mathbf{u}' = A\mathbf{u}$ and therefore $M' = AMA^{-1}$).

A useful way of representing a vector field pictorially is by means of *trajectories*, which are curves for which the vectors of the field are tangent vectors. If the field is a velocity field for a fluid flow in a steady state the trajectories are *streamlines*. The trajectories are the curves determined by the equation

$$\frac{dy}{dx} = \frac{v}{u} \tag{5.51}$$

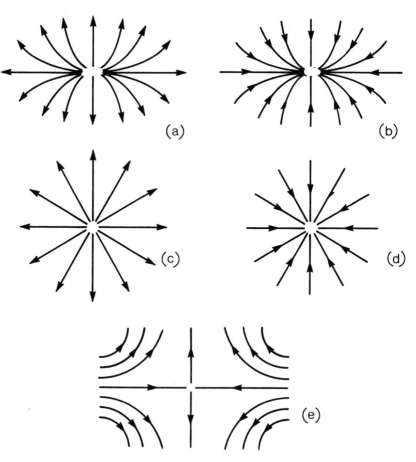

Figure 88

(i) *Both eigenvalues of M real.* The canonical form of the matrix is

$$\begin{pmatrix} \lambda_1 & 0 \\ 0 & \lambda_2 \end{pmatrix}$$

and the trajectories are

$$y = Ax^{\lambda_2/\lambda_1} \tag{5.52}$$

They are shown in Fig. 88, for the cases
(a) λ_1 and λ_2 positive, unequal
(b) λ_1 and λ_2 negative, unequal
(c) λ_1 and λ_2 positive, equal
(d) λ_1 and λ_2 negative, equal
(e) λ_1 and λ_2 of opposite sign.

A multiple of the unit matrix is of course only a special case of the canonical form of a matrix with equal eigenvalues. More generally, a matrix with equal eigenvalues is reducible only to the Jordan canonical form $\begin{pmatrix} \lambda & 0 \\ 1 & \lambda \end{pmatrix}$. The trajectories are given by

$$\frac{dy}{dx} = \frac{x + \lambda y}{\lambda x} \tag{5.53}$$

The curves in the (x, y)-plane connecting points of equal slope, for an equation such as

$$\frac{dy}{dx} = f(x, y) \tag{5.54}$$

are the *isoclines* of the equation [D1]. The isoclines of (5.53) are straight lines through the origin (Fig. 89). This enables us to deduce the appearance

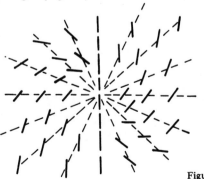

Figure 89

of the trajectories for a vector field in the neighbourhood of a singularity with equal eigenvalues (Fig. 90), of which Fig. 88c is just a special case.

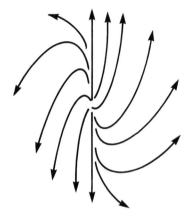

Figure 90

(ii) *Eigenvalues complex conjugate*. The appropriate real canonical form is

$$\begin{pmatrix} \mu \cos \alpha & \mu \sin \alpha \\ -\mu \sin \alpha & \mu \cos \alpha \end{pmatrix} \qquad (5.55)$$

Correspondingly, the trajectories are given by

$$\frac{dy}{dx} = \tan (\theta - \alpha), \quad \frac{y}{x} = \tan \theta \qquad (5.56)$$

Hence the trajectories make a constant angle with the lines through the origin. They are logarithmic spirals. Three cases can be distinguished (Fig. 91):

(a) real parts of roots positive
(b) real parts of roots negative
(c) roots pure imaginary.

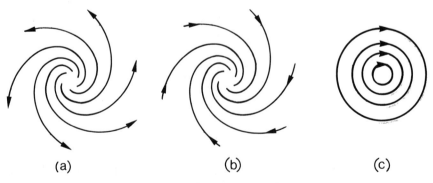

(a)   (b)   (c)

Figure 91

5.12 Vector Field in Three Dimensions

The vector field in three dimensions is describable by a mapping with $M = N = 3$. We denote the coordinates of the space by (x, y, z) and the components of the vector **u** at any point by (u, v, w). At a singular point (chosen to be the origin) we can write a Taylor expansion (5.49), where $\mathbf{u}_n$ is a vector whose components are homeogeneous polymomials of degree n in x, y and z. As for the two-dimensional case, the singularity will be classified according to the eigenvalue structure of M (5.50), which is now a 3×3 matrix.

(a) If the *eigenvalues are all real and unequal*, we have the canonical form

$$\begin{pmatrix} \lambda_1 & & \\ & \lambda_2 & \\ & & \lambda_3 \end{pmatrix}$$

The trajectories are given by $dx : dy : dz = \lambda_1 x : \lambda_2 y : \lambda_3 z$. The parametric descriptons of these curves are

$$\left. \begin{array}{l} x = x_0 e^{\lambda_1 \tau} \\ y = y_0 e^{\lambda_2 \tau} \\ z = z_0 e^{\lambda_3 \tau} \end{array} \right\} \tag{5.57}$$

A pictorial representation of them can be obtained by considering their projections on the three coordinate planes, which produce patterns corresponding to Fig. 88 (a, b or e). This is shown for the case $\lambda_1 = 1, \lambda_2 = 2$, $\lambda_3 = -3$ in Fig. 92. This is a critical point for a *divergenceless* vector field, characterised by the vanishing of the trace of M;

$$\lambda_1 + \lambda_2 + \lambda_3 = 0 \tag{5.58}$$

A complete indication of the field around a singularity with the canonical form (5.57) is given if the field on one coordinate plane, and the set of trajectories on another, are specified. The complete flow pattern is obtained by projection from the plane in which it is given, onto a set of cylindrical surfaces. Thus, in Fig. 93, the complete vector field is obtained by projecting the vectors in the plane $x = 0$ on the hyperbolic surfaces, by projection parallel to the x-axis.

(b) *If two eigenvalues are conjugate complex*, we obtain (for a divergenceless field),

$$\left. \begin{array}{l} u = \alpha x + \beta y \\ v = -\beta x + \alpha y \\ w = -2\alpha z \end{array} \right\} \tag{5.59}$$

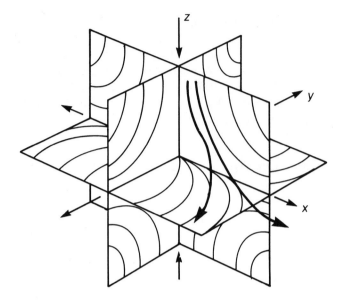

Figure 92

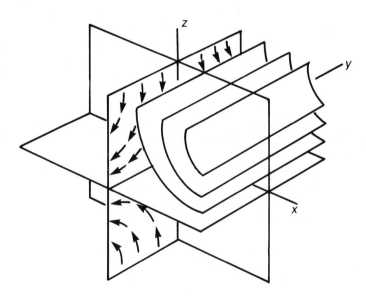

Figure 93

In the plane $z = 0$ we have $w = 0$, and the pattern of trajectories is given by

$$\frac{dy}{dx} = \frac{-\beta x + \alpha y}{\alpha x + \beta y} \tag{5.60}$$

(Fig. 94). Writing $x + iy = \rho e^{i\theta} = \zeta$, and $\alpha + i\beta = \gamma$, the

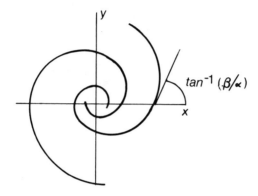

Figure 94

trajectories are

$$\left.\begin{array}{r} \zeta^{\gamma}(\zeta^*)^{-\gamma^*} = \text{const.} \\[4pt] \zeta^{2\alpha} z \gamma^* = \text{const.} \\[4pt] (\zeta^*)^{2\alpha} z^{\gamma} = \text{const.} \end{array}\right\} \tag{5.61}$$

The first equation (5.61) is just $\rho e^{\alpha\theta/\beta} = \text{const.}$, which specifies the cylinders obtained by translating Fig. 92 parallel to the z-axis. Therefore, all the trajectories lie on these surfaces. The other two equations (5.61) give $z\rho^2 = \text{const.}$ and $ze^{-2\alpha\theta/\beta} = \text{const.}$ Thus we have three families of surfaces in which the trajectories lie (Fig. 95). The trajectories are determined as the curves of intersection of any two of these families. An analogue of Fig. 93 is the projection of the vector field in the (x, y)-plane, by projection parallel to the z-axis, on all the surfaces of one of the families $z\rho^2 = \text{const.}$ or $ze^{-2\alpha\theta/\beta}$ (Fig. 96).

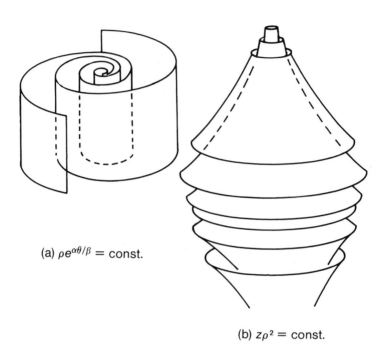

(a) $\rho e^{\alpha\theta/\beta} = $ const.

(b) $z\rho^2 = $ const.

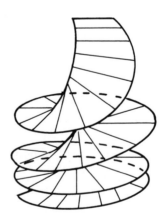

(c) $ze^{-2\alpha\theta/\beta} = $ const.

Figure 95

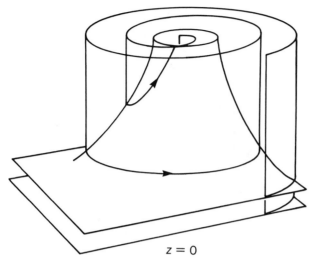

$$z = 0$$

Figure 96

5.13 Boundaries in Fluid Flow

The velocity vector is zero at the surface of any object in a fluid flow. Thus boundaries are surfaces of singularity for the velocity field. If the plane $z = 0$ is a boundary, the field will have the form

$$\left. \begin{aligned} u &= z\xi \\ v &= z\eta \\ w &= z\zeta \end{aligned} \right\} \tag{5.62}$$

and the nature of the flow pattern in the neighbourhood of the boundary is determined by the fictitious 'vector field' with components (ξ, η, ζ). If this field becomes critical at a point on the boundary, one can classify the nature of this critical point as for the critical point of an ordinary vector field.

A crucial indicator of the nature of the flow over a boundary is provided by the two-dimensional vector field

$$\left. \begin{aligned} \xi(x, y, 0) \\ \eta(x, y, 0) \end{aligned} \right\} \tag{5.63}$$

This is the two-dimensional flow that is made visible, for instance, by the drifting of leaves, etc, over the ground. The pattern of air-flow on the windward side of a building is indicated by dotted lines in Fig. 97, which are trajectories for the field (5.63).

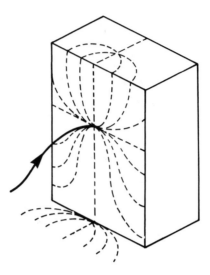

Figure 97

The linear terms in the Taylor expansion of ξ, η, ζ will be written

$$\left.\begin{array}{l} \xi_1 = Ax + Fy + Hz \\ \eta_1 = Ix + By + Dz \\ \zeta_1 = Ex + Gy + Cz \end{array}\right\} \qquad (5.64)$$

If the true field (u, v, w) is divergenceless (corresponding to an incompressible fluid flow), we will have $A + B + 2C = 0$ and $E = G = 0$. The coefficients H and D can be removed by a coordinate transformation affecting only x and y. The classification of these critical points can be based on the eigenvalue structure of the matrix $\begin{pmatrix} A & F \\ I & B \end{pmatrix}$.

(i) *Real roots, opposite sign.* The canonical form (5.64) becomes

$$\left.\begin{array}{l} \xi = x \\ \eta = -\lambda y \; (\lambda > 0) \\ \zeta = \tfrac{1}{2}(\lambda - 1)z \end{array}\right\} \qquad (5.65)$$

The x-axis is a line of separation of a laminar 'boundary layer', for a flow along the y-axis (Fig. 98).

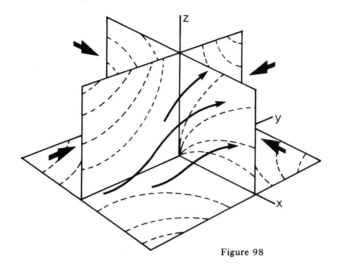

Figure 98

(ii) *Real roots, same sign.* The canonical form is

$$\left.\begin{array}{l} \xi = x \\[4pt] \eta = \lambda y \ (\lambda > 0) \\[4pt] \zeta = -\tfrac{1}{2}(\lambda + 1)z \end{array}\right\} \qquad (5.66)$$

The x-axis is again a line of separation. The difference between this case and the previous one lies essentially in the direction of flow in the surface of separation (the plane $y = 0$), which may be towards or away from the critical point (Fig. 99).

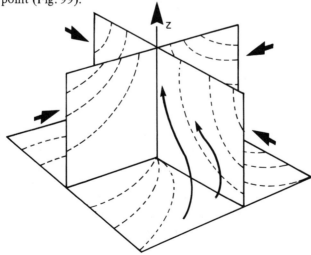

Figure 99

(iii) *Complex roots*. The canonical form is

$$\left. \begin{array}{l} \xi = \alpha x + \beta y \\ \eta = -\beta x + \alpha y \\ \zeta = -\alpha z \end{array} \right\} \qquad (5.67)$$

The pattern somewhat resembles that indicated in Fig. 96, but the families of surfaces are replaced by $\rho e^{\alpha\theta/\beta}$ = const., $z\rho$ = const., $ze^{-\alpha\theta/\beta}$ = const. We have an ascending or descending vortex, Fig. 100.

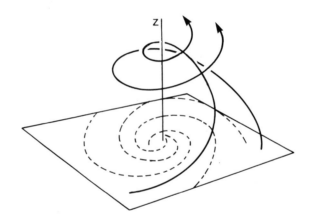

Figure 100

Fig. 101 is taken from a study of flow over a smooth hill, by Hunt and Snyder [H17] Fig. 101a illustrates the streamlines in the vertical plane of symmetry, and Fig. 101b illustrates the trajectories of the surface flow (ξ, η). The various kinds of singularity that we have shown can easily be identified. P_1 is a separation point of the type shown in Fig. 98. P_3 is of the same kind, with the direction reversed. P_2 corresponds to Fig. 99, with the arrows reversed (c.f. Fig. 97). $P_{4(1)}$ and $P_{4(2)}$ are singular points associated with descending vortices (Fig. 100 with the arrows reversed), while P_5 corresponds to Fig. 98, with the arrows reversed.

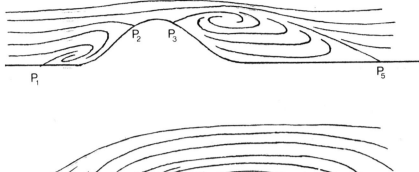

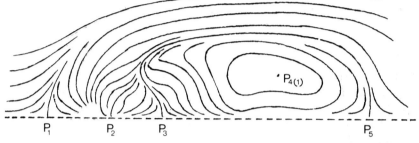

Figure 101

Chapter 6

Mappings that are not analytic

In the previous chapter we have explored some of the special characteristics of forms generated by analytic mappings. However, the requirement of analyticity in mathematical descriptions of form places severe restrictions on the kinds of forms that can be described.

Analytic solutions of the differential equations underlying the physical processes responsible for the shape of an object in the real world (such as a physical field) are at best only approximate descriptions, and are not always obtainable unless the processes involved are particularly simple from a mathematical viewpoint. Forms of objects in the real world are characterised by discontinuities of various kinds. For example, a building presents itself to the physical fields in its environment as a set of surfaces which are associated with abrupt changes in the field variables or their derivatives. The surfaces themselves have edges and corners – another aspect of discontinuity of shape which has to be incorporated in any mathematical description of the interaction.

Even when analytic functions are in principle relevant in a problem of form description, the complexity of the equations governing a process often means that numerical methods have to be resorted to, which, in the finite-element methods [M10], involve an approximate description in terms of non-continuous functions or functions with non-continuous derivatives.

In this chapter we discuss first some of the characteristics of forms generated by mappings, in the neighbourhood of points where the mappings cease to be analytic. We then define the concept of piecewise-analytic mappings, which leads naturally to an investigation of polyhedral shapes.

6.1 Isolated Exception Points
A mapping from M dimensions to N may be analytic except at isolated points. We shall call such a point an isolated *exceptional* point. The possible ways in which this can happen have not been systematically classified in a manner analogous to the classification of critical points for low values of M and N – the situation is much more complicated. We shall therefore simply give a few

examples, for the case $M = 2$, $N = 1$, in order to give an indication of the kinds of situation that can be encountered.

A function $x = f(u, v)$ can become infinite at a point, and yet be analytic at all points in the neighbourhood of this exceptional point. A simple example is

$$x = -\log \rho, \quad \rho^2 = u^2 + v^2 \tag{6.1}$$

illustrated in Fig. 102. The minus sign in (6.1) is included simply to facilitate the pictorial representation.

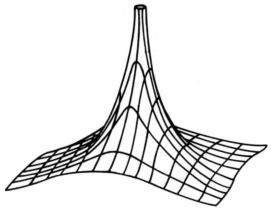

Figure 102

Such a scalar field represents, for example, the potential for a two-dimensional non-viscous fluid flow due to a source or sink, and is related to the singularity at the origin in the conformal mapping ($M = N = 2$) specified by the complex function $w = \log z$ (see §4.6).

The function

$$x = u/\rho^2 \tag{6.2}$$

and its contours are shown in Fig. 103. This is encountered as the potential of a two-dimensional fluid flow due to a dipole, associated with the conformal mapping specified by $w = z^{-1}$.

It is possible for a function $f(u, v)$ to be continuous and finite at a point where its first derivatives become discontinuous. An example of this is given by the functions of the form

$$x = u\alpha(\theta) + v\beta(\theta), \quad \theta = \tan^{-1}(v/u) \tag{6.3}$$

where the functions α and β are analytic and periodic. The general appearance is indicated in Fig. 104. We shall call the exceptional point of a surface which

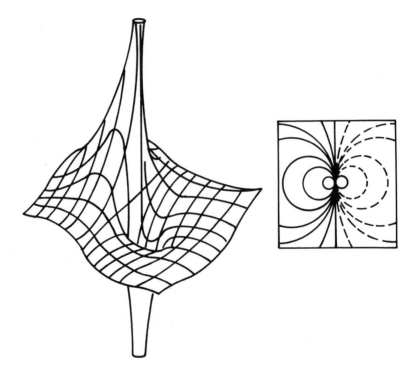

Figure 103

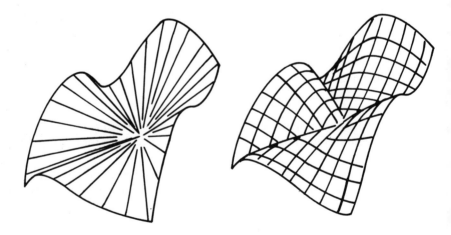

Figure 104

has this kind of form in the infinitesimal neighbourhood of the point a 'conical point'. The tangent vectors for all directions are well-defined at a conical point, but they do not lie in one plane — there is no tangent plane (and no normal) to the surface at such a point.

The surface (Fig. 105) that represents the function

$$x = \sin^2\theta = (v/\rho)^2 \tag{6.4}$$

is a surface with two pinch points. The contours of the functions are straight

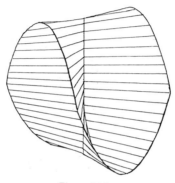

Figure 105

lines through the origin.

6.2 Branch Points

Let $f(u, v)$ be an analytic function defined on some region of parameter space, and let the region be extended by analytic continuation of the function (i.e. by applying the Taylor expansion to define the value of f at new points), until it overlaps itself (Fig. 106). Then, if the two functions defined on the overlapping

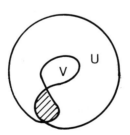

Figure 106

region are not identical, there is necessarily an exceptional point in the region V encircled by the overlapping region U. An exceptional point of this kind is a *branch point*.

Two examples of branch points are the origins for the functions

$$x = \theta \tag{6.5}$$

and

$$x = \rho^{\frac{1}{2}} \cos \tfrac{1}{2} \theta \tag{6.6}$$

(Figs. 107a and b, respectively). In order to keep the function single-valued, a *branch-cut* can be made. This is an artificially introduced line of discontinuity extending from the branch point (but whose position is otherwise arbitrary) (Figs 107c and d).

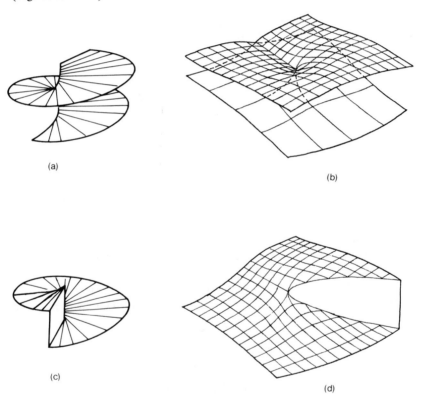

(a)

(b)

(c)

(d)

Figure 107

The exceptional points known as branch points are encountered in the study of analytic complex functions [P7]. The functions (6.5) and (6.6) are the stream function associated with the complex velocity potential $w = \log z$ and the velocity potential associated with $w = z^{\frac{1}{2}}$, respectively.

6.3 Lines of Exceptional Points

An exceptional point of a function may belong to a line of such points (in the neighbourhood of which the function is analytic). We simply indicate the richness of possible situations with the aid of a few examples, without comment (Fig. 108).

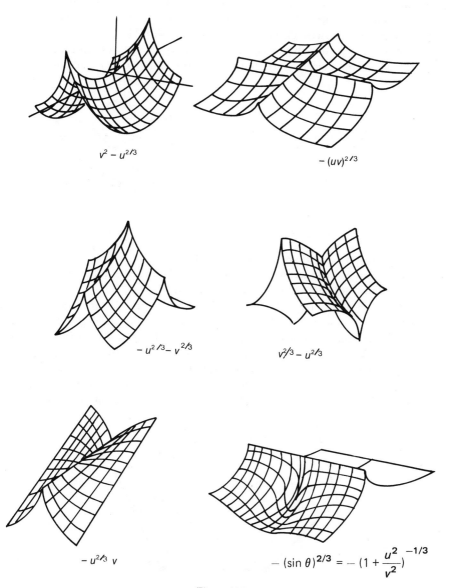

$$v^2 - u^{2/3}$$

$$-(uv)^{2/3}$$

$$-u^{2/3} - v^{2/3}$$

$$v^{2/3} - u^{2/3}$$

$$-u^{2/3}\, v$$

$$-(\sin\theta)^{2/3} = -\left(1 + \frac{u^2}{v^2}\right)^{-1/3}$$

Figure 108

6.4 Edges and Corners

A mapping can be constructed by dividing the image-space into non-overlapping regions and establishing a continuous mapping on each region. We then have a *piecewise continuous* mapping, in general discontinuous on the boundaries between regions. A more useful concept is a *piecewise differentiable* mapping obtained by specifying a differentiable mapping on each of a set of non-overlapping regions of image space, so that the mappings coincide on the boundaries. The images of piecewise differentiable mappings with M = 1, N = 2 (or 3) and with M = 2, N = 3 are polygonal curves and polyhedral surfaces, with the typical appearance of the objects illustrated in Figs. 109 and 110. In a similar way, one

Figure 109   Figure 110

can define mappings that are piecewise analytic.

Important special cases are polygonal arcs and polyhedra in Euclidean space, with rectilinear segments and plane faces, respectively.

The form of a polygonal arc in 3-space is completely specified (apart from its position and orientation) if the lengths of its segments, the angles α between pairs of segments at each vertex, and the dihedral angles β between pairs of osculating planes at each edge, are given. (Equivalently, it is specified by a *sequence of vectors* $s_i t_i$ which give the lengths and directions of its segments.) The osculating plane associated with a vertex is the plane containing the two segments that meet at the vertex (Fig. 111). We shall establish the convention

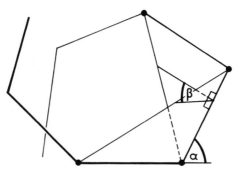

Figure 111

of associating with each segment the osculating plane belonging to the next vertex to be reached. Then every segment except the last has an associated osculating plane, and with every segment except the last we can associate a triad of unit vectors: the tangent vector **t** along the segment, the normal **n** perpendicular to the segment and lying in the osculating plane, and the binormal $\mathbf{b} = \mathbf{t} \times \mathbf{n}$. Then, for two successive segments,

$$\left.\begin{aligned}
\mathbf{t}_2 &= \mathbf{t}_1 \cos \alpha + \mathbf{n}_1 \sin \alpha \\
\mathbf{n}_2 &= -\mathbf{t}_1 \sin \alpha \cos \beta + \mathbf{n}_1 \cos \alpha \cos \beta + \mathbf{b}_1 \sin \beta \\
\mathbf{b}_2 &= \mathbf{t}_1 \sin \alpha \sin \beta - \mathbf{n}_1 \cos \alpha \sin \beta + \mathbf{b}_1 \cos \beta
\end{aligned}\right\} \quad (6.7)$$

These equations are the analogues, for a polygonal arc, of the Serret-Frenet formulae for differentiable curves. In fact, a differentiable curve can be regarded as the limiting case of a polygonal arc as the segment lengths are decreased (and number of segments increased) and the angles α and β are correspondingly decreased. For 'infinitesimally small' angles $d\alpha$ and $d\beta$, the equations (6.7) become

$$\left.\begin{aligned}
d\mathbf{t} &= \mathbf{n} \, d\alpha \\
d\mathbf{n} &= -\mathbf{t} \, d\alpha + \mathbf{b} \, d\beta \\
d\mathbf{b} &= -\mathbf{n} \, d\beta
\end{aligned}\right\} \quad (6.8)$$

Comparison of (6.8) with the Serret-Frenet formulae (2.19, 21, 22) shows that, for a differentiable curve, the *curvature* is the rate of rotation of the tangent vector and the *torsion* is the rate of rotation of the osculating plane, as the curve is traversed at unit speed.

The use of *a piecewise differentiable mapping as an approximation to a differentiable mapping* is of fundamental importance in the development of methods of form description, since piecewise mappings of finite regions with *linear* pieces are completely specified by a *finite number of descriptors*.

6.5 Polyhedra

A *polyhedron* is a closed polyhedral surface immersed in a three-dimensional Euclidean space, whose faces are plane. Topologically, a polyhedron may be homeomorphic to any of the closed surfaces discussed in §3.3. Its faces may intersect, or the perimeter of a face may be a self-intersecting polygon. These situations are illustrated in Fig. 112, which shows one of the four Kepler-Poinsot solids. Its faces are twelve regular pentagrams. Numerous models of 'regular' and 'semi-regular' polyhedra of this general kind have been constructed by Brückner [B11] and, more recently, by Wenninger [W7], their works contain many beautiful photographs of these solids.

A polyhedron is completely described (apart from position and orientation) by a specification of which vertices and edges belong to each face, which edges

Figure 112

and faces meet at each vertex, and which two faces belong to each edge, together with a specification of all edge lengths, facial angles, dihedral angles (i.e. angles between two faces meeting at an edge) and solid angles. Clearly, this description is highly redundant. The question arises, therefore, of what constitutes a *sufficient* description of a polyhedron. That a specification of the positions of all the vertices is not sufficient is evident from the fact that the vertices of the polyhedron shown in Fig. 112 are also those of a regular icosahedron. Nor is a complete description of all the faces, together with the identification of the two faces that share each edge, sufficient. This is evident from Fig. 113, which

 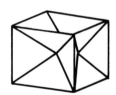

Figure 113

presents two polyhedra which are indistinguishable from the viewpoint of this type of description. What this description does specify, of course, is the *net* of the polyhedron, obtained by cutting along certain edges and folding out the surface onto a plane; the net is the starting point for the construction of a polyhedron from card (Fig. 40). The edge lengths and facial angles of the net of a polyhedron are the analogues of the metric components for a differentiable surface. The dihedral angles and solid angles at vertices (by which the two polyhedra of Fig. 113 can be distinguished) and analogous to the components of the second fundamental form.

The redundancy in the description of the intrinsic geometry of a polyhedron in terms of its edge-lengths and facial angles is evident for a polyhedron whose faces are all triangles (a triangulated polyhedron), since the facial angles are then completely determined by the edge-lengths.

An important descriptor associated with each vertex is the *deficit angle*

$$\varphi = 2\pi - \Sigma\alpha \tag{6.9}$$

where $\Sigma\alpha$ is the sum of all the facial angles that surround the vertex (φ is not necessarily positive).

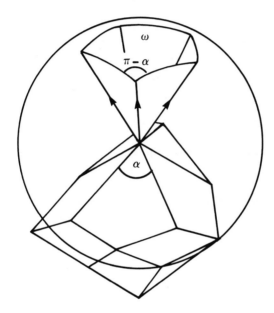

Figure 114

At each vertex, a set of unit normals to the faces that meet at that vertex can be constructed. They define a polygonal cone of directions, whose solid angle will be denoted by ω. The dihedral angles of the cone of directions are the complements of the corresponding facial angles of the polyhedron (Fig. 114). Now, for a triangular cone of directions, $\omega = S - \pi$, where S is the sum of the dihedral angles of the cone (this is just a well-known formula of Gauss for the area of a spherical triangle [C4]). Therefore, since a polygonal cone with n sides can be built up by juxtaposing n triangular cones, we have, for an n-sided polygonal cone, $\omega = S + 2\pi - n\pi$, where S is the sum of the dihedral angles. Therefore, at any vertex of a polyhedron, $\omega = S + (2 - n)\pi = \Sigma(\pi - \alpha) + (2 - n\pi) = 2\pi - \Sigma\alpha$. Hence, at any vertex,

$$\varphi = \omega \tag{6.10}$$

Denoting the solid angle at the vertex of a polyhedron by Ω, and reversing the roles of the vertex and the cone of normals in the above argument, we get $\Omega = 2\pi - \Sigma(\pi - \delta)$, or

$$\Omega = \pi(2 - e) + \Sigma\delta \qquad (6.11)$$

where e is the number of edges meeting at the vertex and $\Sigma\delta$ is the sum of the dihedral angles associated with these edges. Equation (6.11) shows that the solid angles at vertices are not independent parameters. They are completely determined by the dihedral angles.

Consider a polyhedron with V vertices, E edges and F faces. Any face can be triangulated by joining pairs of its vertices; $N - 3$ additional edges partition the face into $N - 3$ triangular faces. By triangulation of all the faces in this way, we obtain a polyhedron with all faces triangular, for which the number

$$\chi = V - E + F \qquad (6.12)$$

is the same as that for the original polyhedron. Also, the sum of all deficit angles is the same. Now, for a polyhedron with triangular faces, every edge belongs to two faces and every face has three edges, so $3F = 2E$. The sum of all the deficit angles of such a polyhedron is $2\pi V$ minus the sum of all the facial angles. But the sum of all the facial angles is just πF. We thus obtain the following expression for the *sum of all the deficit angles* of any polyhedron:

$$\Sigma\varphi = 2\pi(V - E + F) = 2\pi\chi \qquad (6.13)$$

Chapter 7
Minimising principles

When an object is capable of undergoing deformations in response to forces applied to it, it will take up a particular form, which comes about when the applied forces and internal stresses are in a state of balance.

The conditions of balance associated with a stable state of a physical system corresponds to a *minimisation* of the total potential energy of the system. That is, any infinitesimal disturbance of the system from a stable equilibrium configuration is associated with an increase in potential energy. The principle of minimisation of potential energy therefore has an important role in problems of form determination. A very simple example is the spherical shape of a water droplet or bubble, this being the shape that minimises the energy of surface tension.

The principle of least potential energy is the specialisation to static situations of Hamilton's dynamical principle of least action. The appropriate computational technique is the calculus of variations [F4].

In this chapter we illustrate the operation of the minimisation principle by a few examples of the determination of the shapes of curves and surfaces.

7.1 Deformation of a Curve

If a curve in Euclidean 3-space undergoes an infinitesimal deformation, the point with position vector $\mathbf{r}$ is moved to an infinitesimally close point with position vector $\mathbf{r} + \delta\mathbf{r}$. The deformation can be described by referring the infinitesimal vector $\delta\mathbf{r}$ to the triad of reference vectors $\mathbf{t}, \mathbf{n}, \mathbf{b}$ ($\S2.4$):

$$\delta\mathbf{r} = \epsilon\mathbf{t} + \eta\mathbf{n} + \zeta\mathbf{b}. \tag{7.1}$$

By application of the definition $\mathbf{t} = d\mathbf{r}/ds = \dot{\mathbf{r}}$, and the Serret-Frenet formulae for the deformed and undeformed curve, the tangent, normal and binormal, and the curvature and torsion, of the deformed curve, can be computed. We find

$$\left.\begin{array}{l} \delta\,(ds) = \lambda ds \\ \delta t = \gamma n - \beta b \\ \delta n = -\gamma t + \alpha b \\ \delta b = \beta t - \alpha n \\ \delta \kappa = \dot{\gamma} - \kappa \lambda + \tau \beta \\ \delta \tau = \dot{\alpha} - \tau \lambda - \kappa \beta \end{array}\right\} \tag{7.2}$$

where

$$\left.\begin{array}{l} \lambda = \dot{\epsilon} - \kappa \eta \\ \gamma = \dot{\eta} + \epsilon \kappa - \zeta \tau \\ \beta = \zeta + \eta \tau \\ \alpha = \dfrac{1}{\kappa}\,(-\dot{\beta} + \gamma \tau) \end{array}\right\} \tag{7.3}$$

The factor λ represents the amount of stretching undergone by the curve in moving to the new position.

7.2 Fermat's Principle

When a light ray passes through a medium of variable refractive index, the shape of the path of the ray is such that the time taken for the light to travel along it is minimised. That is, for any path infinitesimally close to the actual path, between the same fixed end points, we shall have

$$\delta \int \mu\,ds = 0 \tag{7.4}$$

where μ is the refractive index. Therefore

$$0 = \int \delta \mathbf{r} \cdot (\mathrm{grad}\,\mu)\,ds + \int \mu \delta\,(ds)$$

Without loss of generality, we can take the variation $\delta \mathbf{r}$ to be normal to the ray, so that the parameter ϵ of (7.1) is zero. We then have

$$0 = \int \left[\eta\,(\mathbf{n} \cdot \mathrm{grad}\,\mu - \kappa) + \zeta\,\mathbf{b} \cdot \mathrm{grad}\,\mu\right]\,ds \tag{7.5}$$

Since this must hold for arbitrary infinitesimal parameters η and ζ which are zero at the end-points, the path of a light ray must satisfy

$$\left.\begin{array}{l} \mathbf{n} \cdot \mathrm{grad}\,(\ln \mu) = \kappa \\ \mathbf{b} \cdot \mathrm{grad}\,(\ln \mu) = 0 \end{array}\right\} \tag{7.6}$$

The geometrical content of these equations for a light ray is somewhat clarified if we consider the two-dimensional problem, in which the refractive index μ varies only in the x-direction, and is constant in the y-direction. Equation

(7.6) then gives

$$\kappa = \frac{d\psi}{ds} = -\sin\psi\ \frac{d}{dx}\ln\mu$$

or, since $dx = ds\cos\psi$,

$$\cot\psi\ d\psi = -dx\ \frac{d}{dx}\ln\mu$$

Intergrating, and writing $\psi_1 = i$, $\psi_2 = r$ for the angles that the ray makes with the x-axis at the end-points, and writing μ_1 and μ_2 for the values of the refractive index at the end-points (Fig. 115), we obtain the familiar law of refraction (Snell's law):

$$\frac{\sin i}{\sin r} = \frac{\mu_2}{\mu_1} \tag{7.7}$$

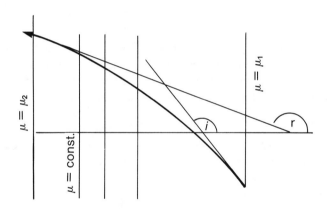

Figure 115

7.3 The Catenary

Perhaps the simplest example of the determination of the form of an object by the forces acting upon it, is that of the form of a hanging chain support at its ends. The potential energy of an element of length ds of a hanging chain of uniform mass per unit length, is proportional to yds. The shape is such that the total potential energy $\int yds$ is minimised. The problem is clearly a two-dimensional problem, so that the parameter ζ in the description (7.1) of small deformations is zero, and without loss of generality we can take also $\epsilon = 0$. Then

$$0 = \delta\int yds = \int (\delta y\ ds + y\delta\ (ds)) = \int (\mu n_y - y\kappa\eta)ds$$

for arbitrary deformations η that vanish at the end-points. Therefore

$$n_y = \cos \psi = y\kappa$$

Divide by κ and differentiate with respect to s:

$$\frac{d}{ds}\left(\frac{\cos \psi}{\kappa}\right) = \frac{dy}{ds} = t_y = \sin \psi$$

Since $\dot{\psi} = \kappa$ (2.7), this becomes

$$-2 \tan \psi = \frac{\dot{\kappa}}{\kappa^2} = \frac{1}{2\kappa^2} \frac{d}{d\psi} \kappa^2$$

This can be integrated immediately to give

$$\kappa = c \cos^2 \psi$$

Integrating again,

$$cs = \tan \psi, \tag{7.8}$$

i.e., the arc-length is proportional to the gradient of the curve. The constant of integration has been eliminated by an appropriate choice of the point from which arc-length is measured. The intrinsic equation for the curve is obtained by eliminating ψ between these last two equations:

$$\kappa = c/(1 + c^2 s^2)$$

Parametric equations for the coordinates of points on the curve are obtained from (2.11). With an appropriate choice of origin, we get

$$\left.\begin{array}{l} cx = \sinh^{-1}(cs) \\ cy = \sqrt{1 + c^2 s^2} \end{array}\right\} \tag{7.9}$$

Finally, the elimination of s between these two equations leads to the familiar form for the equation of the shape of a hanging chain,

$$cy = \cosh cx \tag{7.10}$$

7.4 Flexible Bars and River Meanders

The shape assumed by an elastic bar, when its ends are constrained, is determined by the condition that the energy associated with the deformation shall be a minimum. For the three-dimensional case, we may consider a metal bar of circular cross-section, which is small in relation to its length, and assume that the ends are constrained in such a way that there is no appreciable twisting of the bar. In this case, the energy is proportional to $\int \kappa^2 \, ds$. For an infinitesimal perturbation of the equilibrium shape, which does not move the end points, we find

$$\tfrac{1}{2}\delta \int \kappa^2 \, ds = \int (\kappa\delta\kappa + \tfrac{1}{2}\lambda\kappa^2)\,ds = [\kappa\dot\eta] + \int (\eta(\ddot\kappa + \tfrac{1}{2}\kappa^3 + \tau^2\kappa) - \zeta(\kappa\dot\tau))\,ds,$$

where [ ] denotes the difference between end-point values. Restricting the variations so that the total length of the bar is not varied, we have $0 = \delta\int ds = \int\lambda\,ds = -\int\kappa\eta\,ds$. Hence, the deformed shape must satisfy $\ddot\kappa + \tfrac{1}{2}\kappa^3 + \tau^2\kappa + a\kappa = 0$, $\dot\tau = 0$, and if the direction of the curve at an end point is not constrained ($\dot\eta$ non-zero at an end point), as is the case if a force but not a couple is applied to that end, then $\kappa = 0$ at this position.

For the two dimensional case ($\tau = 0$, $\zeta = 0$), we have the equation satisfied by the shape of a uniform metal strip with constrained ends,

$$\ddot\kappa + \tfrac{1}{2}\kappa^3 + a\kappa = 0 \tag{7.11}$$

Two typical shapes are indicated in Fig. 116.

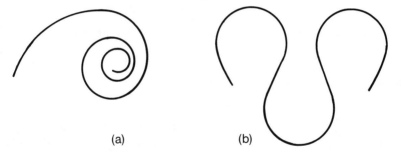

(a)   (b)

Figure 116

The meandering of rivers is also characterised by minimisation of energy associated with bending, and the idealised river meander shape is governed by equation (7.11). If the curvature is not excessive, the κ^3 term can be neglected to obtain a linear approximation to the equation, which has a familiar form, and is satisfied by the 'sine-generated curve'

$$\psi = A \sin \sqrt{a}\, x$$

It was shown by Leopold and Langbein [L5] that well-formed meanders of actual rivers fit these curves quite closely.

7.5 Minimal Surfaces

Plateau's problem [D1, D5, S20] is the problem of determining the shape of the minimum area that is bounded by a given space curve or set of space curves. It is the shape taken up by a membrane over which the tensile stresses are uniformly distributed. The classic example is that of the form of a soap film supported by wires at its boundaries.

The problem is most easily approached by seeking the equation to be satisfied by the function occurring in Monge's form for the description of the surface:

$$z = f(x, y) \tag{7.12}$$

If x and y are taken as the curvilinear coordinates for the surface, the parametric form is

$$\mathbf{r} = \begin{pmatrix} x \\ y \\ f \end{pmatrix} \tag{7.13}$$

It is traditional, when discussing this problem, to introduce the following abbreviations:

$$p = \frac{\partial f}{\partial x}, \quad q = \frac{\partial f}{\partial y}, \quad r = \frac{\partial^2 f}{\partial x^2}, \quad s = \frac{\partial^2 f}{\partial x \partial y}, \quad t = \frac{\partial^2 f}{\partial y^2} \tag{7.14}$$

In terms of these quantities, the metric components are easily found to be

$$E = 1 + p^2, \quad F = pq, \quad G = 1 + q^2 \tag{7.15}$$

The determinant g of the metric is then

$$g = 1 + p^2 + q^2 \tag{7.16}$$

The area of the surface is $\iint \sqrt{g}\, dxdy$, which has to be minimised. For an infinitesimal deformation $\delta f = \epsilon(x, y)$ of the minimal surface, which is zero on the boundaries, we have

$$0 = \delta \iint \sqrt{g}\, dxdy = \iint \delta \sqrt{1 + p^2 + q^2}\, dxdy$$

$$= \iint \frac{p \dfrac{\partial \epsilon}{\partial x} + q \dfrac{\partial \epsilon}{\partial y}}{\sqrt{1 + p^2 + q^2}}\, dxdy$$

$$= -\iint \epsilon \left[\frac{\partial}{\partial x} \left(\frac{p}{\sqrt{1 + p^2 + q^2}} \right) + \frac{\partial}{\partial y} \left(\frac{q}{\sqrt{1 + p^2 + q^2}} \right) \right] dxdy$$

Since this must be valid for *arbitrary* perturbations ϵ, the quantity in square brackets must vanish. We obtain the equation to be satisfied by the function f, in order that the surface (7.12) shall be a minimal surface. The equation is

$$r(1 + q^2) + t(1 + p^2) - 2pqs = 0 \tag{7.17}$$

The meaning of this equation becomes clear if we compute the mean curvature of the surface (7.13). The unit normal is

$$\mathbf{N} = \frac{1}{\sqrt{1 + p^2 + q^2}} \begin{pmatrix} -p \\ -q \\ 1 \end{pmatrix}$$

and the second fundamental form is

$$\begin{pmatrix} D & D' \\ D' & D'' \end{pmatrix} = -\frac{1}{\sqrt{1 + p^2 + q^2}} \begin{pmatrix} r & s \\ s & t \end{pmatrix}$$

Then, from (2.63), the mean curvature is

$$M = -\tfrac{1}{2} (1 + p^2 + q^2)^{-\frac{3}{2}} \; (r (1 + q^2) + t (1 + p^2) - 2pqs) \qquad (7.18)$$

Comparing (7.17) with (7.18), we see that a surface is a minimal surface if, and only if, its mean curvature is zero. This is Plateau's theorem.

Only a very few analytic solutions of equation (7.17) are known. In particular, the helicoid and catenoid discussed in §2.6 have zero mean curvatures and are therefore minimal surfaces. They correspond, of course, to the functions $f(x, y)$ given by $\tan^{-1}(y/x)$ and $\cosh^{-1}\sqrt{x^2 + y^2}$, respectively.

Minimum surfaces are the forms taken by membranes under uniform tension. Numerical solution by the finite element method [M10] is discussed by Ishi [I1].

Chapter 8

Generation of shapes

Shape description is frequently encountered in the form of a set of instructions for the production of a shape. The simplest examples of this are the descriptions of certain plane curves as loci of a variable point. For example, an ellipse can be described as the locus of a point (in a Euclidean plane) which moves in such a way that the sum of its distances from two fixed points remains constant. We shall be concerned in this chapter with a special kind of shape description, consisting of a set of instructions which consist of the iteration of a finite number of simple shape-generating rules.

8.1 Shape Grammars

Shape grammars were developed by Gips and Stiny [G4, S18] as a means of machine-generation of aesthetically pleasing two-dimensional patterns, by the systematic application of simple rules. The theory underlying shape grammars provides insights into the nature of shape and shape description which suggest that the concept may have an importance that goes beyond its use in the production of computer-generated art.

A 'shape' in this context is any set of marks in a Euclidean plane, that can be represented as a drawing on a sheet of paper. The specification of a shape includes the specification of its position and orientation. The *union* and *intersection* of shapes, and the concept of *subshapes* of a shape are defined in the obvious way. The statement that a shape u is a subshape of a shape v is denoted by $u \in v$. The empty shape, or *null shape* is denoted by ϕ. Two shapes are *congruent* if one can be mapped onto the other by a Euclidean transformation, and *similar* if one can be mapped onto the other by a similarity transformation — that is, by a combination of a Euclidean transformation and a dilatation. The image of a shape u under the application of a similarity transformation S is denoted by Su.

A *shape rule* (u, v) is a rule that replaces the shape u by the shape v. It can be applied to any shape w that possesses a subshape similar to u. Specifically, if $Su \in w$,

$$(u, v)w = w - Su + Sv \qquad (8.1)$$

A pair of shape rules R_1 and R_2 are defined in Fig. 117. The initial shape u for each rule appears on the left and the final shape v appears on the right. The dotted line on the right of R_2 does not belong to v, but simply indicates the position of u relative to v. In Fig. 118, we have shown two shapes generated by repeated application of these two rules. Starting from the shape appearing on the left-hand side of R_1, R_1 is repeatedly applied. At each iteration it is applied to all parts of the shape to which it is applicable. After removal of all the dots by applications of R_2, we arrive at the shape shown in Fig. 118a. Five iterations of R_1 will lead to Fig. 118b.

In their analysis of shape grammars, Stiny and Gips make a distinction between *terminals* and *markers*. A marker is a shape which appears in the formulation of the shape rules but which is absent from the generated shapes. The dots in Fig. 117, for example, are markers. A set of shape rules together with a given starting shape constitutes a *shape grammar*. In a *parallel* shape grammar, at each application of a rule, the rule is applied to every part of the shape where it is applicable. The generation of the shapes in Fig. 118, for example, are brought about by the application of a parallel shape grammar.

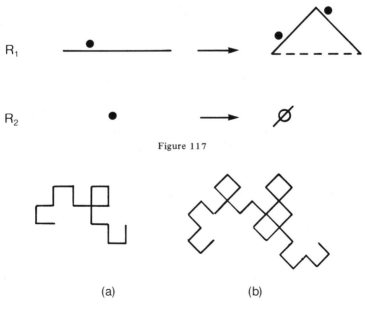

Figure 117

(a)   (b)

Figure 118

The shapes in Fig. 118a and Fig. 118b are 'dragon curves' of fourth and fifth order, respectively. The dragon curves have been discussed by Martin Gardner [G1]. By a modification of the shape rules, the corners of these curves can be 'rounded off', so as to obtain a curve at each iteration which does not touch itself. (The modification involves a more complicated set of shape rules which are not intrinsically interesting). The modified dragon curve of order 10 is illustrated in Fig. 119. The points marked on the figure indicate where the curve can be cut to obtain a dragon curve of lower order. The important conclusion to be deduced from this figure is that apparently intricate and complex configurations can arise as a result of underlying principles of generation that are extremely simple.

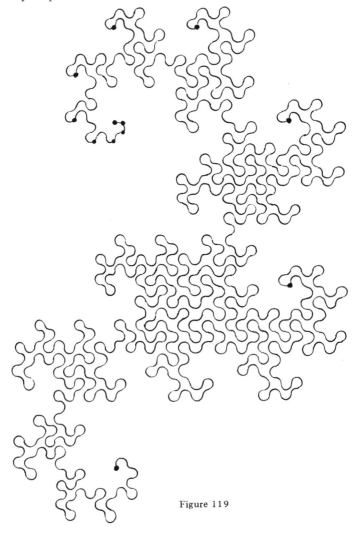

Figure 119

A second example is given by Fig. 120. It leads to patterns with a hierarchical structure, of which Fig. 121 is typical [M1, M2].

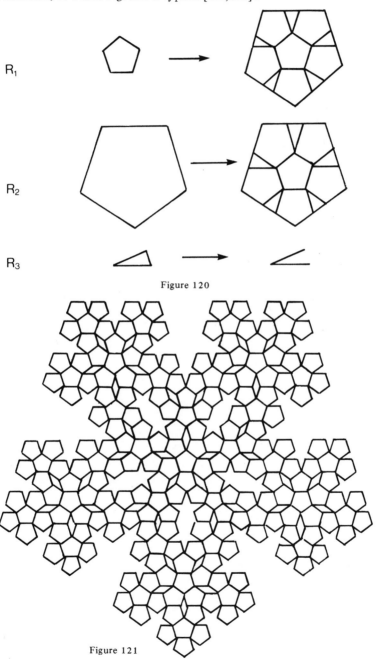

Figure 120

Figure 121

In order to indicate the variety of possibilities, we give three further examples of shape grammars and the shapes they generate. The rule indicated in Fig. 122a, together with a rule for eliminating markers, when applied repeatedly as a parallel shape grammar to the starting shape shown in Fig. 122b, leads to a sequence of curves for which Fig. 123, corresponding to three iterations, is typical. The limiting case (infinite number of iterations) is the pathological curve known as the 'snowflake curve' [K1], which has infinite length but encloses a finite area. The shape grammar defined by Fig. 124 together with a rule for eliminating markers (circles) leads, after two iterations starting from the left-hand side of R_1, to the curve shown in Fig. 125 (the lattices of squares in Fig. 124 are simply an aid to indicate the proportions of the shapes involved). The limiting case for this generating process is Peano's curve [P2], which passes through every point in the interior of a square. Another curve with this property is Hilbert's curve [H1, M12]. A shape grammar for its generation is indicated in Fig. 126a, with the Fig. 126b as the starting shape. The result of three iterations (followed by removal of markers) is shown in Fig. 127. The shape grammars suggested here for these latter two sequences of curves are less complicated than those devised by Gips.

(a)   (b)

Figure 122

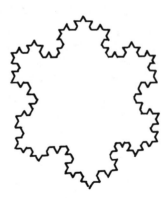

Figure 123

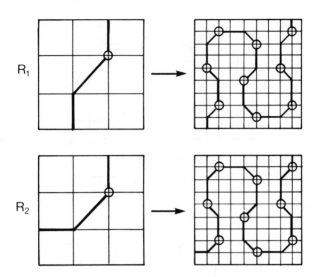

Figure 124

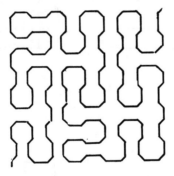

Figure 125

Some aspects of shape grammars in three dimensions, based on simple rectangular blocks and triangular prisms as primitive shapes (specifically, the shapes occurring in Froebel's kindergarten 'gifts') have been investigated by Stiny [S19].

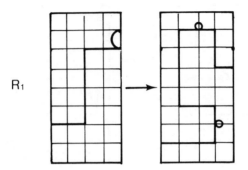

R_1

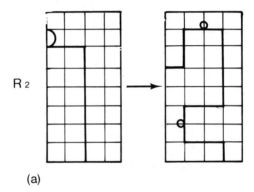

R_2

(a)

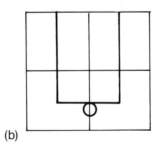

(b)

Figure 126

Figure 127

8.2 Ulam's Modular Patterns

Ulam [S3, U2] has investigated certain kinds of patterns that arise from the accretion of identical subunits; the patterns grow in a manner somewhat similar to the growth of a crystal. The rules for the addition of subunits constitute a shape grammar. The addition of subunits is restricted by *selection rules*, which forbid the addition of a subunit under certain well-defined conditions. The nature of Ulam's patterns is best illustrated by means of examples. Suppose the subunits are identical squares. Starting from a single square (the first generation), let the nth generation consist of squares in contact along a side with the squares of the $(n-1)$th generation. A simple selection rule that forbids the addition of a square that would be in contact with any square (other than its parent) along an edge, will lead, after fourteen generations, to the pattern illustrated in Fig. 128. A similar selection rule forbidding touching at vertices leads to the more open pattern shown in Fig. 129.

Of course, more complicated selection rules will lead to still other patterns. Ulam has investigated patterns based on triangular and hexagonal lattices, as well as those, like Fig. 128 and Fig. 129, based on a square lattice, and has also investigated the consequences of rules whereby generations disappear after a certain number of iterations. The surprising thing about these patterns is the complexity that can arise from quite simple rules of growth.

Although all of Ulam's patterns were based on lattice arrangements, this is not strictly necessary. Fig. 130 shows a pattern based on a pentagonal subunit,

Figure 128

Figure 129

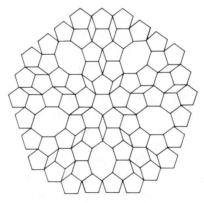

Figure 130

with a simple selection rule forbidding a subunit to touch, along a side, two subunits of the previous generation.

The generalisation of Ulam's patterns to three dimensions, in which the subunits are cubes, has been investigated by Schrandt and Ulam [S3].

8.3 Trees and River Systems

The growth of a tree is an example, in nature, of the generation of a shape by the action of simple rules. Therefore some understanding of the characteristics of form that distinguish different species of tree should be possible by means of shape grammars, at least at the level of idealisation that ignores the environmental influences that lead to variations of form within a single species. Such an approach was discussed by Stevens [S16]. A typical parallel shape grammar that leads to a bifurcating branched structure is defined by the rule given in Fig. 131. The shape obtained after four iterations is shown in Fig. 132. The resemblance between this figure and the pattern of branching of a tree is quite striking. Variations of proportion in Fig. 131 will lead to a branching pattern belonging to a different 'species'. The reader is referred to Stevens for a more detailed discussion.

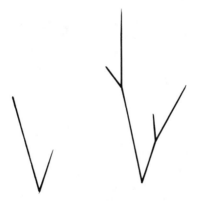

Figure 131

Another obvious example of a branched form in nature is that of a river and its tributaries. In this case, the form is generated by processes that involve an element of randomness, corresponding to the random variations in the terrain over which the river system flows. The numerical simulation of branching structures based on random application of simple rules was investigated by

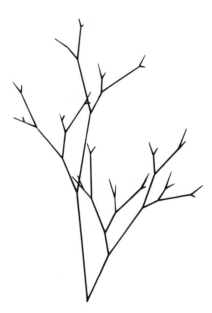

Figure 132

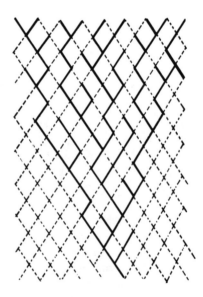

Figure 133

Leopold and Langbein [L3], and their work is discussed in Stevens' *Patterns in Nature* [S16]. A branching pattern on a lattice is created by a random choice of two possible 'directions of flow' at each point (Fig. 133). That is, we apply a shape grammar consisting of a pair of shape rules, in which the decision about which rule to apply at each application is a random process. A statistical analysis of the branched forms generated in this way reveals certain form characteristics which approximate to those of real river systems [H12]. Defining a first order branch to be one with no tributaries, and an Nth order branch to be one served only by tributaries of order less than N, the Nth order branches are found to be four times more numerous than $(N + 1)$th order branches, for real river systems as well as for the randomly-generated patterns. Moreover, for real river systems as well as for the random patterns, the length of the main channel and the area covered by the pattern are related by

$$L = \kappa A^{\nu} \tag{8.2}$$

with ν between ·6 and ·7.

8.4 Symmetry

If a form in the Euclidean plane, or in Euclidean 3-space, is mapped onto itself by the action of any transformation belonging to a subgroup G of the Euclidean group, then G is the *symmetry group* of the form.

A form with a discrete symmetry group G can be regarded as a shape generated by applying the shape grammar consisting of a set of shape rules

$$u \to u + gu, \tag{8.3}$$

where the g are transformations belonging to G, to a *subunit u*.

The *finite* subgroups of the plane Euclidean group are C_N, generated by rotation through an angle $2\pi/N$ about a fixed point (the *centre of symmetry*), and D_N, generated by such a rotation and a reflection in an axis through the symmetry centre. For example, the symmetry group of a swastika is C_4, that of a snowflake is D_6, (Fig. 134), and that of any object with *bilateral symmetry* is $D_1 = m$. The fact that the two sequences C_N and D_N give all possible symmetries of finite objects in a plane was recognised by Leonardo da Vinci, in connection with the classification of building plans.

It is convenient to represent a symmetry group by a *regular point system*, which is the result of applying the shape grammar (8.3) to a single point. If the chosen point is not in any special relationship to the transformations (such as lying on a reflection axis or at a symmetry centre), the regular point system will be a *general* one for the particular group. For example, Fig. 135 shows the appearance of general regular point systems for C_5 and D_5.

The classification of the possible symmetries of finite objects in Euclidean 3-space involves an investigation of the finite subgroups of the three-dimensional Euclidean group. These groups leave a fixed point of Euclidean space invariant

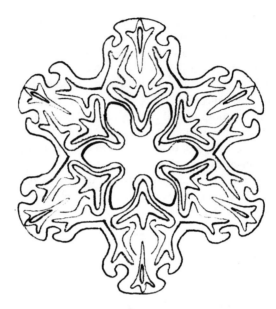

Figure 134

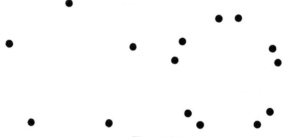

Figure 135

(the centre of symmetry), and hence are called *point groups*. They are the discrete subgroups of the group of rotations of a sphere. The associated regular point systems are sets of points on the surface of a sphere. A complete list of the point groups is as follows:

(a) The symmetry groups of the regular prisms, and their subgroups.

(b) The symmetry groups of the regular antiprisms, and their subgroups.

(c) The symmetry group of the regular icosahedron, and its subgroups.

A regular pentagonal prism, a regular pentagonal antiprism, and a regular icosahedron, are illustrated in Fig. 136. The notation we shall use for the various groups is taken from Coxeter [C14]. The symmetry group of a regular N-gonal prism is $D_N \times I$ (I denotes inversion in the symmetry centre) if N is even and $D_{2N}D_N$ if N is odd. The symmetry group of an N-gonal antiprism is $D_N \times I$

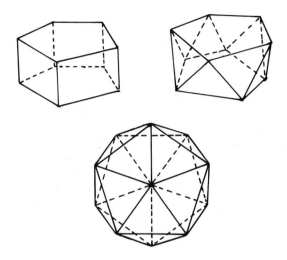

Figure 136

if N is odd and $D_{2N}D_N$ if N is even. The action of these groups, and their sub-groups (for N = 5 or 6), is indicated in Fig. 137 by means of the general regular point systems associated with them. Note that the group D_N in this context contains only rotations (no reflections) and is generated by a rotation $2\pi/N$ and a rotation π, the two rotation axes being orthogonal. The symbol $C_{2N}C_N$ means that the action of C_{2N} is modified by combining the transformations of its subgroup C_N with an inversion; $D_{2N}D_N$ and D_NC_N are defined similarly.

The symmetry group of the icosahedron and those of its subgroups that are not already included in the list of groups and subgroups for prisms and anti-prisms, are as follows. The notation is that of Coxeter [C14].

(a) A_5. The rotational group of the icosahedron, of order 60. That this group is necessarily a subgroup of the permutation group S_5 is seen by noting that the 20 mid-points of faces of an icosahedron are the vertices of a cluster of five cubes, which are permuted by rotations of the icosahedron. One such cube is indicated in Fig. 138.

(b) $A_5 \times I$. The symmetry group of the icosahedron, order 120.

(c) S_4. The rotational group of a cube, order 24. It is isomorphic to the group of permutations of four objects. (The four main diagonals of the cube are permuted by the action of the group.)

(d) $S_4 \times I$. The symmetry group of the cube, order 48.

(e) A_4. The rotation group of the regular tetrahedron, order 12.

(f) S_4. The symmetry group of the tetrahedron, order 24.

(g) S_4A_4. A subgroup of the symmetry group of the cube, of order 24.

The appearance of the general regular point-sets associated with these groups is shown in Fig. 139.

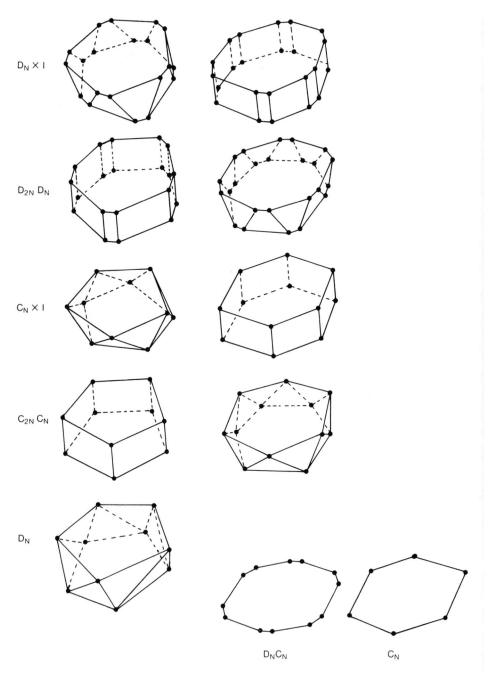

Figure 137

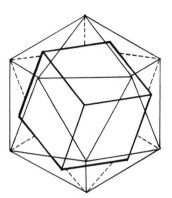

Figure 138

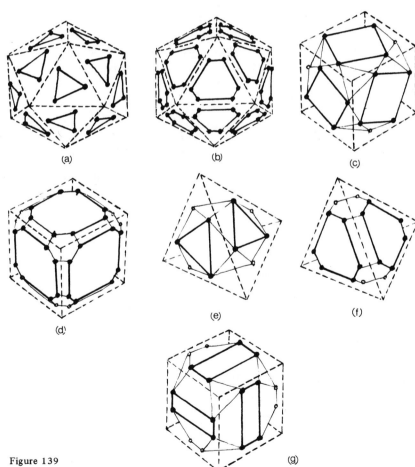

(a)   (b)   (c)

(d)   (e)   (f)

Figure 139   (g)

8.5 Regular Patterns in a Plane

A shape with a discrete symmetry group that contains two independent *trans-lations* will extend over the whole plane, forming a repeating pattern. There are just 17 such discrete subgroups of the two-dimensional Euclidean group, which are sometimes given the name 'wallpaper groups'. A *lattice* in a Euclidean plane is a regular point system generated by two translations. The lattices can be classified according to the point group that leaves a lattice point fixed and maps the lattice into itself; these are just five kinds of two-dimensional lattice, illustrated in Fig. 140 for each possible point symmetry. The two cases that arise for D_2 are distinguished according to whether the translations are orthogonal to each other or not. It follows, from this classification of lattices, that a point group can be a subgroup of a wallpaper group only if it is a subgroup of D_6 or D_4. This enables the wallpaper groups to be classified and enumerated. We give below, for each wallpaper group, the symbol used to denote it in the International Tables for X-ray Crystallography, the symbol given to it by Fejes Tóth [F1], and a set of transformations for the shape rules (8.3) that will generate patterns with the given symmetry group. A *glide reflection* is a combination of a reflection and a translation parallel to the reflection axis (Fig. 141).

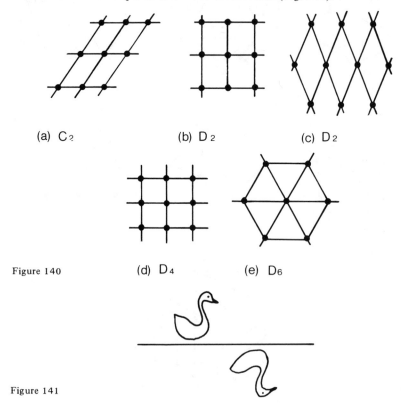

(a) C_2 (b) D_2 (c) D_2

Figure 140   (d) D_4 (e) D_6

Figure 141

1. $p1, W_1$.   Two translations.
2. $p2, W_2$.   Two translations and a rotation π (half-turn).
3. cm, W_1^1.   A reflection and a parallel glide reflection.
4. pm, W_1^2.   Two perpendicular translations and a reflection with axis parallel to one of them.
5. pg, W_1^3.   A translation and a perpendicular glide reflection.
6. cmm, W_2^1.   Two perpendicular reflections and a half-turn.
7. pmm, W_2^2.   Two perpendicular reflections, and a translation parallel to each of them.
8. pmg, W_2^3.   A reflection, a translation parallel to it, and a half turn. Or, two perpendicular translations, a reflection parallel to one and a glide-reflection parallel to the other.
9. pgg, W_2^4.   Two perpendicular glide-reflections.
10. $p3, W_3$.   A translation and a rotation $2\pi/3$.
11*. $p3m1, W_3^1$.   A reflection in three side of an equilateral triangle.
12*. $p31m, W_3^2$.   A reflection and a rotation $2\pi/3$ *not* centred on the reflection axis.
13. $p4, W_4$.   A translation and a rotation $\pi/2$.
14. $p4m, W_4^1$.   A translation, a parallel reflection, and a rotation $\pi/2$ centred on the reflection axis.
15. $p4g, W_4^2$.   A glide reflection and a rotation $\pi/2$.
16. $p6, W_6$.   A translation and a rotation $\pi/3$.
17. $p6m, W_6^1$.   A reflection and a rotation $\pi/3$, not centred on the axis.

The general point systems for these groups are illustrated in Fig. 142. The lines have been inserted for ease of visualisation, and have no particular significance.

The possibilities for the decoration of plane surfaces by the regular repetition of a motif, are infinite. Owen Jones' *Grammar of Ornament* [J3] contains a very large number of examples, from many cultures. Fig. 143a (Chinese) and Fig. 143b (Egyptian) are taken from this work. The exploration of regular geometrical patterns to be seen in Islamic decorative art [B5, C17, E5, J3] is particularly remarkable. A discussion of the classification of housing scheme layouts in terms of the wallpaper groups is given by March and Steadman [M3].

The best available introduction to the elementary concepts of symmetry is probably Weyl's book [W8], which is lavishly illustrated with photographs and drawings of man-made and natural forms possessing symmetry properties.

*The notation here corrects the misprint in C14 (p. 413), whereby these two groups were interchanged (see C15).

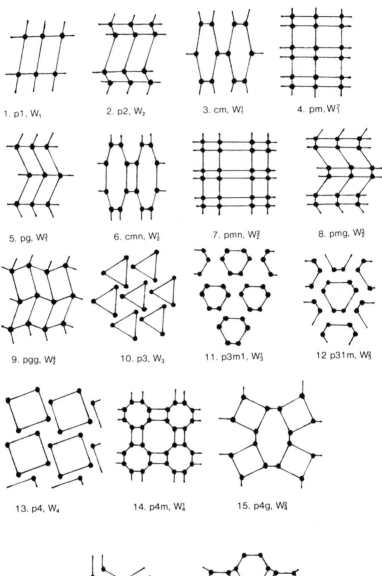

1. p1, W_1 2. p2, W_2 3. cm, W_1^1 4. pm, W_1^2

5. pg, W_1^3 6. cmn, W_2^1 7. pmn, W_2^2 8. pmg, W_2^3

9. pgg, W_2^4 10. p3, W_3 11. p3m1, W_3^1 12 p31m, W_3^2

13. p4, W_4 14. p4m, W_4^1 15. p4g, W_4^2

16. p6, W_6 17. p6m, W_6^1

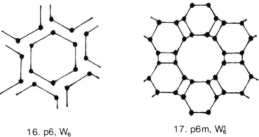

Figure 142

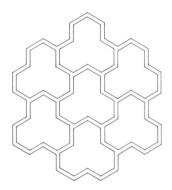

Figure 143

8.6 Crystallography

A *lattice* in Euclidean 3-space is a regular point system generated by three translations. A *primitive cell* of a lattice is a parallelepiped with a point at each vertex, but not in its interior or on any face or edge. A *symmetry* of a lattice, that keeps *one point fixed*, is necessarily one of the following seven point groups, so any lattice belongs to one of *seven crystal systems*:

cubic: the cubic group, $S_4 \times I$.
hexagonal: symmetry group of a hexagonal prism, $D_6 \times I$.
tetragonal: symmetry group of a square prism, $D_4 \times I$.
rhombohedral or trigonal: symmetry group of a triangular antiprism, $D_3 \times I$.
orthorhombic: symmetry group of a 'digonal prism', $D_2 \times I$.
monoclinic: symmetry group of a 'monogonal prism', $D_1 \times I$ (generated by a
    reflection and an inversion centred on the reflection plane).
triclinic: group of order 2 generated by an inversion, I.

The number of distinct lattice types in 3 dimensions is fourteen. A portion of each of the *fourteen Bravais lattices* is illustrated in Fig. 144. The letters P, B, I, F stand for 'primitive', 'base-centred', 'body-centred' (*innenzentriert*) and 'face-centred'. To clarify the three-dimensional situation, a cube or a regular hexagonal prism accompanies some of the figures. An example of a primitive cell has been marked on the base-centred orthorhombic lattice (dotted lines). Many different choices of primitive cell exist for each Bravais lattice.

A *space group* is the 3-dimensional analogue of a wallpaper group. It is a discrete subgroup of the 3-dimensional Euclidean group that contains three independent translations. A shape grammar (8.3), based on a space group, applied to a subunit *u*, will generate a three-dimensional pattern which extends over the whole of the space. Such patterns occur in nature in the arrangement of molecules in a crystal. A point group can be a subgroup of a space group if,

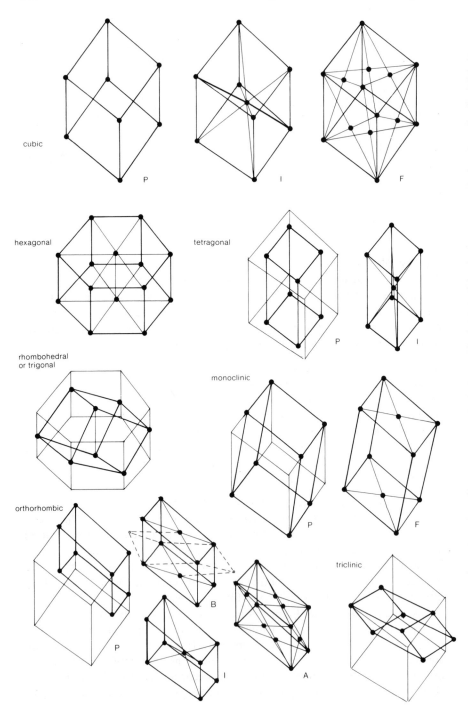

cubic

hexagonal   tetragonal

rhombohedral
or trigonal

monoclinic

orthorhombic

triclinic

Figure 144

and only if, it is one of the seven groups listed above, or a subgroup of one of these groups. There are altogether 32 such point groups; this gives rise to the classification of crystals into 32 *crystal classes*.

The classification and enumeration of space groups involves the investigation of which crystal classes are consistent with each of the Bravais lattices, when an object with the symmetry of a point group is placed at each lattice point. The classification is complicated because of the special care needed to avoid counting a point group twice, and extra complication comes from the fact that a reflection belonging to a point group may be realised, in a space group, as a *glide reflection* (combined reflection and translation parallel to the reflection plane), and a rotation belonging to a point group may be realised as a 'rotatory translation' (combined rotation and translation along the rotation axis) along a *screw axis*. A very clear description of the systematic enumeration of space groups is given by Phillips [P8].

Seven possible kinds of screw axis are illustrated in Fig. 145 by means of the regular point systems associated with rotatory translations. There are four more, namely $3_2, 4_3, 6_4$ and 6_5, which are the mirror images, respectively, of $3_1, 4_1, 6_2$ and 6_1.

From the point of view of abstract group theory, the number of space groups is 219. But, because of the phenomenon of screw axes, eleven of these groups are realisable as patterns in three-space in two different ways which are mirror images of each other. This is responsible for the phenomena of enantiomorphic (mirror-image) pairs of crystals, and of optical activity in crystalline materials. Hence the total number of space groups, taking enantiomorphic pairs into account, is 230.

An understanding of the *external form* of a well-formed crystal is based only on the 32 crystal classes. The external form of a crystal consists of several sets of similar faces. Each set is referred to as a '*form*'. For example, three such 'forms' can be identified in the crystal illustrated in Fig. 146; the planes that constitute each form are the faces of a cube, an octahedron, and a rhombic dodecahedron. All the faces of a single form are generated from *one* face by the application (as a shape grammar) of the point group of the crystal class to which the crystal belongs.

A description of the shape of a crystal is obtained by representing the normals, from the centre of symmetry to the faces of the crystal, by points on a sphere. Each form is then represented by a regular point system on the sphere, which can be projected on a plane by *stereographic projection* (§4.4). The crystal shown in Fig. 146 is then represented by Fig. 147. The cube corresponds to the black dots, the octahedron to the white dots, and the rhombic dodecahedron to the double circles. When representing a single 'form' in this way, it is conventional to represent the points within the diametral circle (broken line in Fig. 147) by black dots and those outside it by white dots (small circles) at their inversive images in the diametral circle. The eight points of the octahedron,

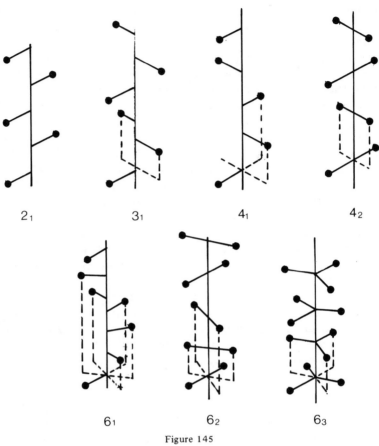

2_1 3_1 4_1 4_2

6_1 6_2 6_3

Figure 145

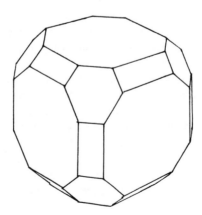

Figure 146

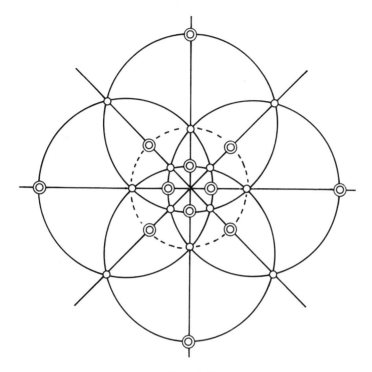

Figure 147

according to this convention, would be indicated by Fig. 148. This is the method used by crystallographers to denote diagrammatically the regular point systems generated by the point groups of the crystal classes [P8] .

A coordinate system in Euclidean 3-space can be related to a Bravais lattice so that the points of the lattice have integer coordinates. A plane of lattice points will then intercept the axes at $\alpha/h, \alpha/k, \alpha/l$, where h, k and l are *intergers*. Thus, a family of parallel planes containing lattice points is specified by the ratio of a set of three integers h, k, l. Thus the 'form' generated from one such plane by applying the point group of the crystal can be referred to as the form $\{h, k, l\}$. For example, the cube, octahedron and rhombic dodecahedron of the crystal shown in Fig. 146 (which has the full cubic symmetry $S_4 \times I$) are the forms $\{100\}$, $\{111\}$ and $\{110\}$, respectively. These are special forms belonging to the cubic class. The *general* forms belonging to the cubic group consist of 48 planes. They are 'hexoctahedra' like Fig. 149.

For the hexagonal system, it is usual to employ four coordinate axes, related to the lattice as indicated in Fig. 150. Planes and forms are then specified by the ratios of four integers $\{hkil\}$ satisfying $h + k + i = 0$.

If, instead of using stereographic projection to map the spherical representa-

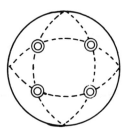

Figure 148

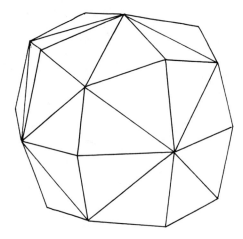

Figure 149

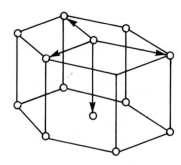

Figure 150

tion of the crystal to a plane, we use gnomonic projection (ie; projection from the centre of the sphere), then the Miller indices *hkl* are simply homogeneous projective coordinates for the plane. Rotations of the crystal correspond to a projective transformation in the plane. This representation is sometimes used, but is less common than the stereographic representation.

Only a few of the infinite number of possible 'forms' associated with a given crystal class will manifest themselves as sets of faces on any given crystal. Which 'forms' will occur is dependent on crystal *habit*, which is concerned with the different growth rates of the various 'forms'. Thus, the shapes of crystals of different substances belonging to the same class, or the shapes of crystals of the same substance grown under different conditions, may be quite different in appearance because of variations in habit. Thus the description of form involved when a crystal is assigned to one of the 32 classes is a particularly clear illustration of the way in which a method of form description singles out particular aspects of form, while discarding other aspects as irrelevant to the purpose of the description; the purpose of the description by classification, in this instance, being the need to gain some insight into the internal arrangement of molecules in a given crystal.

Mathematical crystallography is an excellent illustration of the way in which the properties of three-dimensional Euclidean space constains the possibilities for the morphological characteristics of objects. The seven crystal systems, 32 crystal classes and 230 space groups form the totality of traditional mathematical crystallography. Mackay [M1, M2] has argued for the generalisation and extension of the scope of this subject matter to embrace more general aspects of morphology, by considering not only perfectly 'regular' repetition of subunits, but also repetition according to other well-defined rules. Ulam's systematically generated patterns [S3, U2] and patterns with hierarchical structure such as Fig. 121 were cited by McKay as examples. Other possible extensions of the concept of discrete symmetry will be indicated in the following two sections.

The space group to which a crystalline solid belongs gives, of course, only a very incomplete indication of the internal structure at a molecular level. A complete description of the atomic arrangement within a crystal, provided by the techniques of X-ray crystallography, consists of a specification of the positions of all the atoms in a primitive cell of the Bravais lattice. The patterns revealed can be extremely intricate, and are generally highly systematic in a way that goes beyond the systematics of the space group. We shall give just one example to clarify this statement. Boron can form a compound with a metal, in which the ratio of boron atoms to metal atoms is 1 : 12, which belongs to the face-centred cubic system, and in which the boron atoms are arranged at the vertices of a space-filling of *truncated octahedra* and *cuboctahedra*, while the metal atoms are at the centres of these polyhedra [S21] (Fig. 151).

A useful introduction to the atomic structure of crystals is Battey's *Miner-*

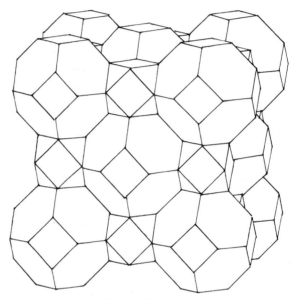

Figure 151

alogy for Students [B1]. Very thorough treatments of the subject are given in Bragg's *Atomic Structure in Minerals* [B6] and in Bragg and Claringbull's *Crystal Structure of Minerals* [B7]. An entertaining non-technical discussion is given by Loeb [L9].

8.7 Tessellations and Space Filling

A plane *tessellation* is a covering of the Euclidean plane by non-overlapping regions ('tiles').

Tessellations which have a 'wallpaper group' as a symmetry group are of special interest. The so-called 'regular' and 'semi-regular' patterns are of this kind; they are defined by requiring all the vertices (of the network formed by the tile edges) to be equivalent under the group (so they form a regular point system) and by further requiring the tiles to be regular polygons. There are just three regular tessellations: the familiar coverings of the plane by squares, equilateral triangles, or regular hexagons ('honeycomb' pattern). There are eight semi-regular patterns (more than one tile shape), of which Fig. 152 is an example [F1, C16]. It is denoted by the symbol $(4, 3^2, 4, 3)$, which specifies the order in which the tiles are clustered around each vertex. The symmetry group of the pattern is p4 (§8.4). The other seven semi-regular patterns are $(3, 12^2)$, $(4, 6, 12)$, $(4, 8^2)$, $(3, 4, 6, 4)$, $(3, 6, 3, 6)$, $(3^4, 6)$, $(3^3, 4^2)$. The symbol $(3^4, 6)$ has actually two realisations, which are mirror images.

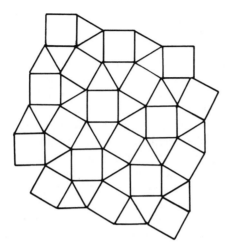

Figure 152

Analogous considerations of tessellations of the surface of a sphere lead to the five Platonic solids (tetrahedron, cube, octahedron, icosahedron, dodecahedron) and the thirteen Archimedean solids, and the regular prisms and antiprisms.

Alternatively, one can consider tessellations with a 'wallpaper' symmetry in which the tiles themselves, rather than the vertices, are equivalent under the group (and therefore congruent). For example, the *duals* of the eight semi-regular tessellations (obtained from perpendicular bisection of their edges) are tessellations of this kind. Fig. 153, for example is the dual of $(4, 3^2, 4, 3)$. The construction of duals of semi-regular tessellations is a particular instance of the construction of the Dirichlet, or Voronoi, regions for a set of points in the plane [C14]. A Dirichlet region for a point P in a plane, belonging to a set of

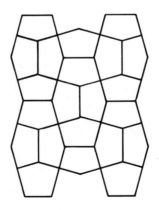

Figure 153

points, is a polygon with the property that every point in its interior is closer to P than to any other point of the set. The Dirichlet regions of a set of points in a plane form a tessellation of the plane. The network formed by their edges is the medial axis (§5.7) of the point-set; if we imagine circular wave-fronts spreading outwards from each point of the set, with uniform velocity, the loci of points of intersection of wavefronts will be edges of Dirichlet regions. In particular, every *regular* point system has a unique associated tessellation of the plane into congruent, convex polyhedra, namely the tesselation whose tiles are the Dirichlet regions. For example, the tessellation associated with the general regular point set generated by the group p2 (Fig. 142) is indicated in Fig. 154.

The three-dimensional generalisation of the above concepts leads to the consideration of space-filling by polyhedra [C16, P4]. If the pattern is required to have one of the 230 space groups as a symmetry group, and if all vertices are to be equivalent, there are only four solutions with congruent regular or semi-regular solids (cube, truncated octahedron (Fig. 155), triangular prism, hexagonal

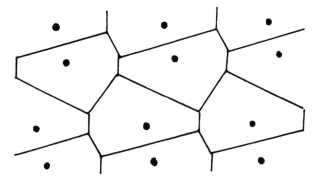

Figure 154

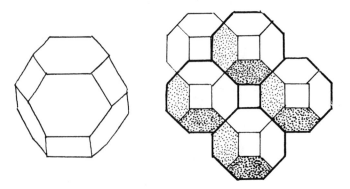

Figure 155

prism). If more than one kind of regular or semi-regular solid is allowed, there are 19 more solutions [C16, P4]. Fig. 151 is one of them.

Every regular point system in 3-space is associated with a space-filling by congruent polyhedra. For example the *rhombic dodecahedron* (Fig. 156) is a Dirichlet region for the face-centred cubic Bravais lattice, and Critchlow's space-filling solid [C16] (Fig. 157) is the Dirichlet region for the regular point-set corresponding to the arrangement of atoms in a diamond.

The concept of *symmetry* can be generalised by considering the discrete subgroups of continuous groups of transformations other than the Euclidean

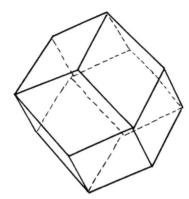

Figure 156

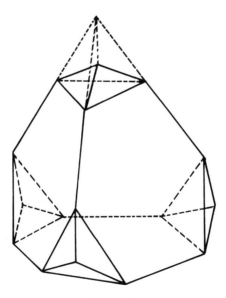

Figure 157

group. We obtain corresponding generalisations of the concept of regular point systems, and of tessellations. Fig. 158, for example shows a portion of a tessellation of the interior of a circle, whose 'symmetry' group is a subgroup of the inversive group (§2.9). From the point of view of this group, the tiles are all 'congruent'. The figure is a representation, in the Euclidean plane, of a tessellation of the hyperbolic plane [C11]. The work of the Dutch artist Escher [C15, L8] contains many designs based on such generalised concepts of symmetry, as well as many ingenious tessellations with conventional Euclidean symmetries.

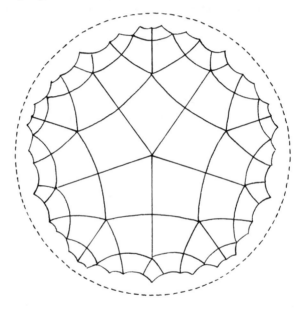

Figure 158

A different kind of generalisation comes from the investigation of the possibility of tessellating the Euclidean plane with one or more sets of congruent tiles, without requiring the pattern to have a 'wallpaper' symmetry. These are called 'aperiodic' tessellations. Fig. 159 is a simplified version of a pattern with a single tile shape, devised by Voderberg [F1, G2, V5], which has only the point symmetry C_2. It can be extended, in a spiral arrangement, to cover the whole plane. More intricate tessellations with spiral structure have been given by Grünbaum and Shephard, in The Mathematical Gardner [K4].

R. Penrose [P9] has devised pairs of polygons which, when used in conjunction as tiles, produce infinitely many aperiodic tessellations, which in general are 'random', but may in particular cases possess point symmetries or 'hierarchical' structure. The Penrose tiling patterns are discussed by Martin Gardner [G2]. As an example, we illustrate the central portion of a Penrose pattern which has

Figure 159

D_5 symmetry (Fig. 160). Other aspects of aperiodic tessellations are discussed by Schattenschneider [S1] and by Grünbaum and Shephard [G8].

A different kind of generalisation comes from considering adjacent units arranged in a systematic manner (according to rules based on a space group, or in some other systematic way) without filling the space. Fig. 121 and Ulam's patterns are examples in the plane. Open clusters of polyhedra, arranged according to space groups, have been extensively investigated by Pearce [P3, P4]. A hierarchical clustering of icosahedra, suggested by Pearce, is a three-dimensional analogue of Fig. 121. If twelve icosahedra are arranged around a central icosahedron (in contact with each face), the outer planes of the cluster are the planes of a larger icosahedron. Hence a 'supercluster' of 13 of these clusters can be formed, and the process repeated indefinitely (Fig. 161).

8.8 Spiral Phyllotaxis

In the previous section, we mentioned the possibility of extensions of the concept of symmetry, based on groups of transformations other than the Euclidean group. An obvious extension is the enlargement of the Euclidean group to include *dilatations*.

The most general kind of similarity transformation in the Euclidean plane is a *spiral similarity*, which is a combination of a rotation and a dilatation with the same centre. A transformation of this kind, together with its inverse, generates a one-dimensional subgroup of the group of similarities. The appearance of a

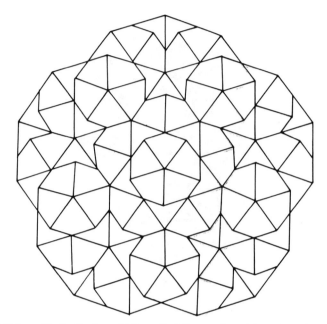

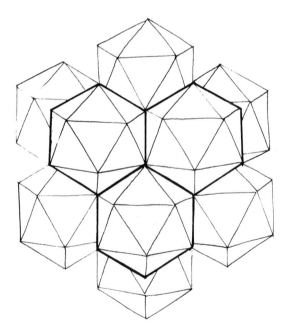

Figure 161

regular point system associated with such a one-dimensional group is shown in
Fig. 162. There is a single logarithmic or equiangular spiral on which all the
points of the system lie. In botanical terminology, this is the *ontogenetic* spiral.
Two important descriptors of a pattern generated by a shape rule based on a
spiral similarity are the angle of this equiangular spiral, and the angle subtended
at the centre by two successive points (i.e. the angle of the spiral similarity). As
is apparent from looking at Fig. 162, there are other equiangular spirals containing sets of points of the system. For example, there are three congruent spirals
containing respectively the points $(3, 6, 9. . .)$, $(4, 7, 10 . . .)$ and $(5, 8, 11 . . .)$,
and five congruent spirals containing $(5, 10, 15, . . .)$, $(6, 11, . . .)$, $. . . (9, 14, . . .)$.
Hence a pattern generated by a shape rule based on a single spiral similarity can
also be generated (in many ways) by a pair of shape rules based on two independent spiral similarities (Fig. 163).

The system shown in Fig. 162 is characteristic of the arrangement of buds
(leaf or flower primordia) around the tip of a growing plant stem, which gives
rise to the arrangement of leaves on the stem, or the arrangement of flowers
in an inflorescence. The arrangement comes about because the angle subtended
at the centre of the tip by two successive primordia remains constant. The spirals
that are most apparent to the eye in a pattern of this kind are the *parastichies*,
which are curves connecting *spatially adjacent* subunits. Fig. 164a shows a
pattern with the salient parastichy numbers 3 and 5, and Fig. 164b shows one
with parastichy numbers 8 and 13 (a parastichy number $m - n$ means that the
mth primordium is adjacent to the nth). In botanical terminology, these patterns
are said to exhibit $(3, 5)$-phyllotaxis and $(8, 13)$-phyllotaxis respectively. Both
patterns are generated by the *same* spiral similarity; parastichy numbers are
determined by the shape of the subunit as well as by the underlying spiral
similarity. Parastichies are particularly evident in a pineapple, a pine-cone and

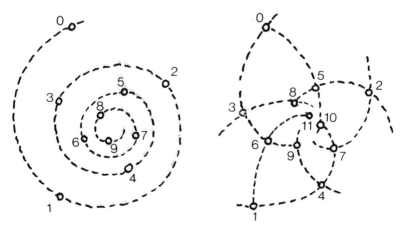

Figure 162   Figure 163

Figure 164

a sunflower. The inflorescence of a cauliflower exhibits a remarkable hierarchical structure. The general appearance is something like Fig. 164a, where each subunit is a cluster of subunits arranged in the *same* pattern, and each second-order subunit is again an arrangement, in the same pattern, of the individual florets.

The two simplest phyllotactic systems are the *distichous* and *decussate* systems, illustrated in Fig. 165, which indicates, for each system, typical cross-sections of the stem apex and the primordia surrounding it. The distichous pattern is generated by a combination of rotation through an angle π combined with dilatation. The decussate system is generated by rotation through $\pi/2$ combined with dilatation, but the subunit is a pair of diametrically opposed primordia which are created simultaneously. Clearly, distichous phyllotaxis can lead to an arrangement of leaves on a stem describable by a 2_1 screw axis, and decussate phyllotaxis can give rise to a 4_2 screw axis — both of which are common forms.

The parastichy numbers found in spiral plant forms are usually found to be consecutive terms of the *Fibonacci sequence* $1, 1, 2, 3, 5, \ldots$ (in which each term is the sum of the two preceding terms). In the sunflower, Fibonacci phyllotaxis as high as $(89, 144)$ has been observed. This is a remarkable fact that calls for explanation. We shall briefly indicate the theory put forward by van Iterson [V2] and developed by Snow [S10]. The theory is based on the hypothesis that each primordium is formed in the largest gap available in the pattern formed by the already existing primordia. This implies that the second primordium will be diametrically opposite the first. In Fig. 166, the stem apex has been mapped on a plane by representing angles on a horizontal axis and the logarithm of

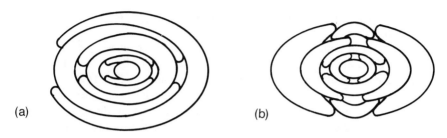

(a)   (b)

Figure 165

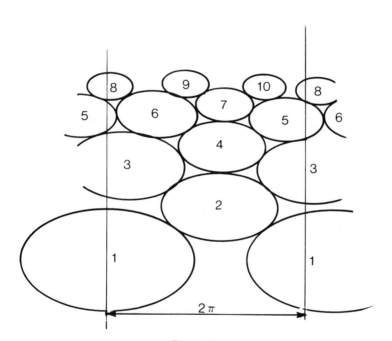

Figure 166

distances from the centre on a vertical axis. The first few primordia, with cres-
cent-like shapes as in Fig. 165a, will appear as ovals in this representation (the
exact shape is immaterial). Because of decreasing available space, we suppose a
gradual diminution of the angular extent of primordia. Thus, in the figure,
the fifth primordium is not diametrically opposite the third; the initial distichous
system breaks down and becomes $(1, 2)$-phyllotaxis. A further transition occurs
with the formation of the seventh primordium, which makes contact with the
fourth. The system is then $(2, 3)$-phyllotaxis. Transition to $(3, 5)$-phyllotaxis
occurs at the tenth primordium. Thus we see how spiral phyllotaxis with gradu-

ally increasing Fibonacci parastichy numbers can originate from an initially distichous sytem.

The ratio of consecutive terms of the Fibonacci sequence $(\frac{1}{2}, \frac{2}{3}, \frac{3}{5}, \frac{5}{8}, \ldots)$ tends, in the limit, to the 'golden ratio' $\tau = \frac{1}{2}(\sqrt{5} - 1) = 0.618 \ldots$. It follows that, if the angle subtended by two successive points in Fig. 162 is $2\pi\tau$ (or, equivalently, $2\pi(1 - \tau) = 137°30'$ approximately), the radii associated with points labelled by Fibonacci numbers lie increasingly close to the radius through the point marked 0. The mode of development indicated by Fig. 166 will lead, after a 'steady state' with high parastichy numbers has been reached, to a pattern closely resembling an 'idealised' pattern in which the angle subtended at the centre by successive primordia is close to the special Fibonacci angle. The angle for actual plant forms exhibiting spiral phyllotaxis is found to be quite close to the Fibonacci angle. In fact, it tends to be much closer to the Fibonacci angle than the above theory would lead us to expect, in the case of low parastichy numbers such as $(1, 2)$ or $(2, 3)$. This suggests that the zones indicated in Fig. 166 should not be regarded as the shapes of the actual visible primordia, but as zones of influence that serve to determine their position; as we have already pointed out, the parastichy numbers are dependent on the shape of subunit, so it is possible for the pattern associated with the mechanism determining the manner of growth to have higher parastichy numbers than those of the visible pattern of actual primordia.

As well as the work of van Iterson and Snow already cited, interesting discussions of spiral phyllotaxis have been given by Richards [R2, R3], Wardlaw [W4, W5], Coxeter [C14] and Stevens [S16].

8.9 Spiral Forms in Three Dimensions

A spiral similarity in Euclidean 3-space is a combination of a rotation and a dilatation whose centre is on the rotation axis. If the Cartesian coordinate system is chosen so that the centre of the spiral similarity is the origin and the axis of the similarity is the z-axis, the matrix of the transformation has the form

$$\rho \begin{pmatrix} \cos\alpha & \sin\alpha & . \\ -\sin\alpha & \cos\alpha & . \\ . & . & 1 \end{pmatrix}. \tag{8.4}$$

A spiral similarity generates a one-dimensional discrete subgroup of the Euclidean group. The shape grammar (8.3) based on such a group, applied to the point P whose position vector is $\mathbf{a}$, gives rise to a regular point system whose points have the position vectors

$$\mathbf{r}_n = \rho^n \begin{pmatrix} \cos n\alpha & \sin n\alpha & . \\ -\sin n\alpha & \cos n\alpha & . \\ . & . & 1 \end{pmatrix} \mathbf{a}, \tag{8.5}$$

where n is any positive or negative integer, or zero. These points all lie on a *concho-spiral* whose parametric equation is

$$\mathbf{r}(v) = e^{kv} \begin{pmatrix} \cos v & \cos v & . \\ -\sin v & \sin v & . \\ . & . & 1 \end{pmatrix} \mathbf{a}, \qquad (8.6)$$

where

$$k = \tfrac{1}{2} \ln \rho. \qquad (8.7)$$

This curve is obtained by projecting the logarithmic spiral $r = e^{k\theta}$ in the (x, y) plane on the cone generated by rotating the line OP about the z-axis, by parallel projection along the z-axis.

Applying the shape grammar to a space curve $\mathbf{r} = \mathbf{a}(u)$ gives rise to a system of similar space curves

$$\mathbf{r}_n(u) = \rho^n \begin{pmatrix} \cos n\alpha & \sin n\alpha & . \\ -\sin n\alpha & \cos n\alpha & . \\ . & . & 1 \end{pmatrix} \mathbf{a}(u), \qquad (8.8)$$

which all lie on the *surface*

$$\mathbf{r}(u, v) = e^{kv} \begin{pmatrix} \cos v & \sin v & . \\ -\sin v & \cos v & . \\ . & . & 1 \end{pmatrix} \mathbf{a}(u). \qquad (8.8)$$

The surface can be regarded as the result of repeated application of an *infinitesimal* spiral similarity (let $\rho \to 1$, $\alpha \to 0$, but keep k, defined by (8.7), fixed) to a given curve $\mathbf{a}(u)$. Such a surface is completely specified by the 'generator' $\mathbf{a}(u)$ and the number k. The parametric curves $u =$ const. are concho-spirals and the parametric curves $v =$ const. are curves similar to the generator. *Any* curve on the surface that is not parallel to a v-curve at any point, can be used as a generator.

Typical surfaces of this kind, with the parametric curves $u =$ const, $v =$ const. marked on them, are illustrated in Fig. 167.

Surfaces of this kind are characteristic of forms which grow by a process of accretion, in which material is deposited along a space curve (the generator) which grows but retains its shape, the rate of deposition at any point on the generator being proportional to the rate of growth of the generator. The shells of molluscs and the horns of ruminants are very striking examples in nature of this kind of growth [T3].

8.10 Generation of Surfaces
The surfaces discussed in the previous section are examples of surface generation by repeated application of an infinitesimal similarity transformation to a space curve. Other examples are the *cylinders* (generated by translation), the *cones*

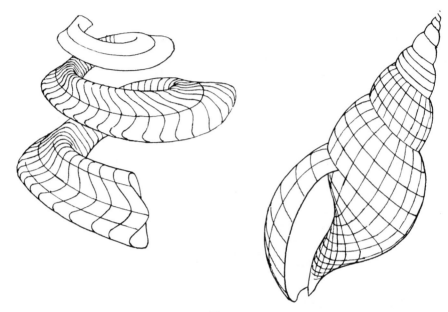

Figure 167

Figure 168

(generated by dilatation) and the *surfaces of revolution* (generated by rotation). When the generator is a straight line, the surface of revolution is a *hyperboloid* ('of one sheet') (Fig. 168). It is a ruled surface containing two families of generators.

The helicoid, discussed in §2.5, is a particular example of a surface generated by a *rotatory translation*. The generator is a straight line. Fig. 169 illustrates a surface obtained by applying, to a circle, a rotatory translation whose axis is perpendicular to the plane of the circle. It has been employed for ornamental

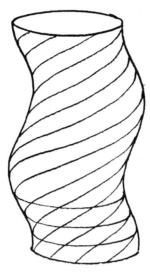

Figure 169

effect in the shapes of columns, in Baroque architecture. Particularly fine examples are to be seen in Bernini's baldachino in St. Peter's, Rome [C4] .

The *ruled* surfaces [E1] are often employed in architectural and civil engineering forms because of the simplification of construction entailed by the use of straight lines. Any ruled surface is specified by a pair of space curves and a continuous one-to-one mapping between them. No generality is lost by taking the two curves to be plane sections of the surface. For example, the hyperboloid in Fig. 168 is completely specified by a mapping between two circles, and is obtained by joining pairs of corresponding points by straight lines. If $\mathbf{r} = \mathbf{f}(u)$ and $\mathbf{r} = \mathbf{g}(u)$ are the parametric equations of the two curves, with the parametrisations chosen so that pairs of points that correspond under the mapping have the same value for the parameter, then a parametric equation for the associated ruled surface is

$$\mathbf{r} = v\mathbf{f}(u) + (1 - v)\mathbf{g}(u) \qquad (8.10)$$

A particularly simple ruled surface is the *hyperbolic paraboloid*, which, in the versions used most frequently in roof construction, is obtained by joining pairs of points spaced uniformly along a pair of straight lines (Fig. 170). Like the hyperboloid of revolution, the hyperbolic paraboloid contains two families of straight lines. A particular hyperbolic paraboloid was encountered already in §5.2 (Figs. 56 and 57), as the surface $z = x^2 - y^2$ (or $z = xy$). The hyperbolic paraboloid and the hyperboloid of one sheet are examples of ruled *quadric* surfaces [C7, H8] .

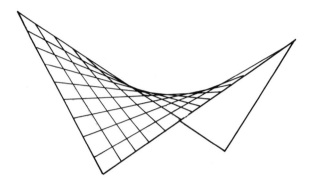

Figure 170

A different kind of example of the generation of special surfaces by simple rules is provided by the *translation surfaces*, which are obtained by sliding a space curve along another, keeping its orientation fixed and a chosen point of the moving curve always in contact with the stationary curve. Interchanging the roles of the two curves will give the same surface. If the two curves are $\mathbf{r} = \mathbf{f}(u)$ and $\mathbf{r} = \mathbf{g}(v)$, a parametric equation for the resulting surface is simply

$$\mathbf{r}(u, v) = \mathbf{f}(u) + \mathbf{g}(v) \tag{8.11}$$

Fig. 171 indicates how the 'twisted column' of Fig. 170 can be regarded as a translation surface based on a circle and a helix whose axis is perpendicular to the plane of the circle.

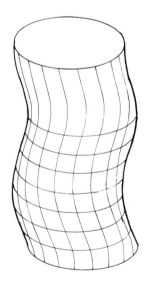

Figure 171

The hyperbolic paraboloid is a translation surface obtained from two parabolae with oppositely oriented axes (Fig. 172).

If the two curves defining a translation surface are polygonal curves, the resulting surface will be a polyhedral surface whose faces are parallelograms (Fig. 173). Such a surface is specified by a pair of ordered sets of vectors, each

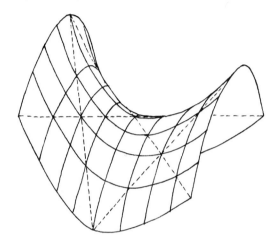

Figure 172

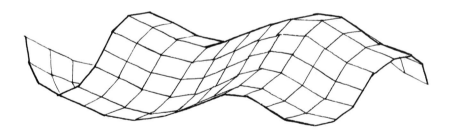

Figure 173

of which gives the lengths and directions of the line segments of one of the curves. In general, there are four parallelogram faces at each vertex of the surface (a vertex around which the number of faces is different from four, as in Fig. 173b, is analogous to a singularity on a differentiable translation surface). Polygonal surfaces whose faces are all parallelograms are called 'zonohedra'. Some of their properties have been investigated by Homegraaf [H11].

Fig. 174 illustrates a roof structure designed by Eduardo Catalano [K2] consisting of twelve hyperbolic paraboloids. This example is included to indicate the design possibilities revealed by studying the generation of surfaces by the application of systematic rules to a basic subunit.

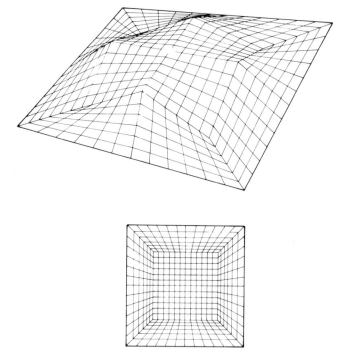

Figure 174

A survey of architectural structural forms, which extends the concepts indicated briefly in this section, is provided by Engel's *Structure Systems* [E5]. See also Siegel's *Structure and Form in Modern Architecture* [S8] and *Structure in Art and Science*, edited by Kepes [K2].

Chapter 9

Discrete spaces

Up to this point, the geometrical spaces that have provided the basis for the study of shapes have been *continuous* spaces. The shapes themselves have for the most part also been characterised by the possession of *continuity* properties (e.g. continuous curves and continuous surfaces). However, in the previous chapter, shapes were encountered which were devoid of continuity properties, namely, the regular point systems, which consist entirely of discrete points, which could be finite or infinite in number. These *discrete* shapes were nevertheless conceived as objects immersed in continuous spaces.

Certain aspects of a regular point system are independent of the immersion of the system in a continuous space. If the points of a regular point system are labelled by assigning letters or numbers to them in some arbitrary way, the action of the symmetry group of the system can be described as a group of *permutations* of the labels. This group of permutations of a set of symbols is clearly an aspect of the regular point system that can be studied without reference to its interpretation as a symmetry group of an object in a continuous space. From this point of view, the point system is itself a geometrical space, and the permutation group is its *group of motions* (§2.9).

9.1 Finite Spaces
Let S be a set of N objects ('points') and let G be a subgroup of the group of permutations of these objects, which acts *transitively* — that is, there is a permutation belonging to G that will map any point to any other. Then S is a *finite geometrical space* and G is its *group of motions*.

In this way, Klein's Erlangen program (§2.9) is extended to include the concept of geometries consisting of only a finite number of points. There is a geometry associated with every transitive subgroup of a permutation group.

For example, consider a set of five points, labelled by the symbols 12345. Consider a subgroup D_5 of the group of permutations of five labels. One such subgroup is generated by the cyclic permutation (12345) and the permutation (24)(35). The corresponding five-point geometry contains five 'triangles' 123,

234, 345, 451, 512 which are equivalent under the group (the analogue of 'congruency' in this finite geometry). The remaining triangles of the geometry are 135, 241, 352, 413, 524, and these are also equivalent to each other, but not equivalent to any triangles of the first set. As for continuous geometries, a transformation can be regarded 'actively' or 'passively'. In the active interpretation a permutation specifies a mapping of a set of points onto its image; in the passive interpretation it specifies a *relabelling* of the points. There exist mappings of the five-point geometry to a Euclidean plane with the property that the permutations are realisable as Euclidean transformations; the image of the five points are the vertices of a regular pentagon and the action of the permutations in the finite space correspond to the action of the symmetry group D_6 of the pentagon.

The finite projective geometries have particularly interested mathematicians. They have been extensive studied [A2, V1, V3]. A particularly simple example is the geometry of seven points and seven lines with three points on each line and three lines through each point (indicated schematically in Fig. 175), where the representation of one of the lines has been drawn as a circle. This emphasises the fact that the configuration is not realisable in a Euclidean plane). The corresponding projective group is a subgroup, of order 168, of the group of permutations of seven labels, which permutes the triplets of labels indicated in the figure.

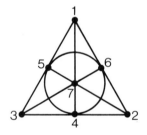

Figure 175

The concept of finite spaces leads naturally to the idea of discrete spaces with an infinite number of discrete points. An example is a two-dimensional *lattice* space, whose points are labelled by pairs of *integers* (i, j) (positive, negative, or zero), and whose group of motions consists of the 'translations' $(i, j) \rightarrow (i + k, j + l)$, where k and l are integers. This geometry is obviously obtainable by abstracting the permutational properties of the general lattice in a Euclidean plane. Every infinite regular point set has an associated infinite discrete geometry obtained by abstraction of permutational properties in this way.

9.2 Graph Theory

An aspect of the theory of finite spaces, of wide application, is *graph theory* [O1, O2]. A *graph* is a form in a finite space, defined by distinguishing *point pairs*. A distinguished point-pair (i, j) is called an *edge* of the graph. If we distinguish *ordered* point pairs, the graph is a *directed* graph. If, for each distinguished pair, (i, j) is not regarded as distinct from (j, i), the graph is *undirected*. *Mixed* graphs can also be considered.

Two graphs with N vertices each are *isomorphic* if they can be mapped onto each other, i.e. if there is a one-to-one correspondence between the points of the two graphs such that every edge of one graph corresponds to an edge of the other. A convenient diagrammatic representation of a graph consists of a figure in which edges are indicated by joining its points by a line (not necessarily straight). A directed edge can be labelled by an arrow in the diagram.

Isomorphism is not always readily apparent from the diagrammatic representations of a pair of graphs. For example, the isomorphism of the two graphs indicated in Fig. 176 can be recognised only by verifying that the mapping between the two point-sets, indicated by the cyclic permutation (265734), is an isomorphism.

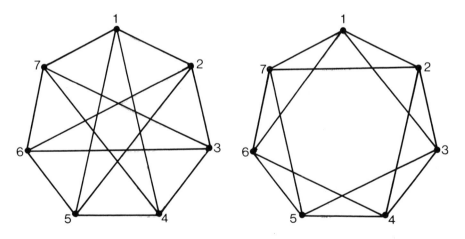

Figure 176

There are various ways in which graph theory could be generalised. One could define forms in a space of N points by distinguishing triples, rather than pairs of points (as in Fig. 175), for example. Or one could allow the number of points of a graph to be infinite, with edges defined in some systematic way.

9.3 Planar Graphs

A *planar graph* is one which can be represented by a diagram on a plane, in which the lines representing edges do not intersect.

A criterion for deciding whether or not a given graph is planar is provided by the theorem due to Kuratowski. First note that any graph which contains points belonging to only one edge or only two edges can be associated with a *contracted* graph, obtained from it by removal of those points. Fig. 177, for example, represents a graph and its associated contraction. It is obvious that the question of whether or not a graph is planar is the same, for the graph and for its contracted form. Hence only graphs in which every point is associated with three or more edges need be considered. A graph is *planar* if, and only if, its contracted form does not contain one of the graphs shown in Fig. 178. This is *Kuratowski's* theorem.

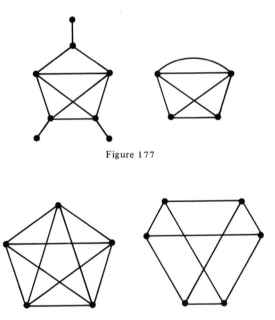

Figure 177

Figure 178

Every planar graph can be represented by a diagram in a plane, consisting of vertices, non-intersecting edges, and faces. If the diagram is drawn on the surface of a sphere, it can be regarded as a tessellation of the surface (the exterior region of the planar diagram becomes an extra 'tile', or face). With every such representation is associated a *dual* representation, obtained by interchanging faces and vertices of the spherical representation. For example, Fig. 179a represents a planar graph, and Fig. 179c is the dual representation.

 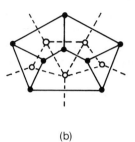 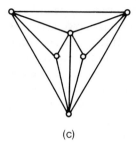

(a)   (b)   (c)

Figure 179

This aspect of graph theory has been applied to the problem of enumerating the possible arrangements of rooms in a single storey of a building. The plan of a storey can be regarded as a diagram of a planar graph, whose edges are walls and whose vertices are conjunctions of three or more walls. For example, the room layout indicated in Fig. 180 is topologically equivalent to Fig. 179a (the rooms need not be rectangular and the walls need not be straight). A complete description of the contiguity properties of the layout (which rooms are adjacent, and which have external walls) are completely represented by Fig. 179c. The problem of enumerating the topologically distinct configurations for N rooms reduces to the problem of enumerating the planar graphs with N + 1 vertices. This problem is dealt with, for example, by Steadman [S14], who gives the complete set of solutions up to N = 6, and by March and Earl [M6], Baybars and Eastman [B2], etc.

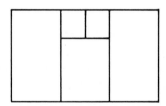

Figure 180

9.4 Euler's Formula

We have seen that every planar graph can be represented as a tessellation of the surface of a sphere. Clearly, only *topological* properties of the tessellation are relevant to the description of the graph. Non-planar graphs can be represented as tessellations of closed surfaces which are not homeomorphic to spheres.

Consider a tessellation of a closed surface, with V vertices, E edges and F faces (the faces having the simple topology of 'discs'). By making cuts along the edges of the tessellation, the surface can be mapped to the interior of a plane polygon, as described in §3.2. The internal edges and vertices of the tessellation of this polygon can be successively removed by applying the following rules:

(i) If a vertex has only one edge connected to it, remove the vertex and edge.

(ii) If the two faces adjoining an edge are distinct, remove the edge. This decreases the number of faces by 1.

(iii) If a vertex has only two edges connected to it, remove it and replace the two edges by a single edge.

Eventually, only the external edges and vertices will remain. At each stage, the number

$$\chi = V - E + F \tag{9.1}$$

does not change. The polygon that remains reconstitutes the surface when pairs of edges are identified in the manner discussed in §3.3. The cutting and rejoining of this polygon involved in its reduction to canonical form do not change the number χ. Hence, the number χ for any tessellation of a closed surface is the same as the number χ for the fundamental polygon specified by the polygonal symbol (3.2) or (3.3) of the surface. Therefore, χ is a topological invariant of the surface, and *the same for any tessellation of the surface*. It is called the *Euler characteristic* of the surface.

Of course, the fundamental polygon described by the canonical form (3.2) or (3.3) provides a tessellation of the surface with only *one* face. For a sphere, the polygonal symbol AA^{-1} refers to a tessellation with two vertices, one edge and one face. Therefore $\chi = 2$ for a sphere. For an orientable surface with genus $g(> 0)$, the symbol (3.2) refers to a tessellation with one vertex, $2g$ edges, and one face. Therefore the Euler characteristic for a closed *orientable surface of genus g* is

$$\chi = 2 - 2g \tag{9.2}$$

For a non-orientable surface with k cross-caps, the symbol (3.3) refers to a tessellation with one vertex, k edges, and one face. The Euler characterstic for a closed *surface with k cross-caps* is therefore

$$\chi = 2 - k \tag{9.3}$$

Any continuous, differentiable surface in Euclidean 3-space can be approximated by a polyhedral surface. Such a polyhedral surface can be obtained by constructing the tangent planes at numerous points on it. For a closed surface, the closed polyhedral approximation to it satisfies (6.10) and (6.13), or

$$\Sigma\omega = 2\pi\chi \tag{9.4}$$

In the limit, the left-hand side can be replaced by an integral over the surface, which then leads, on applying equation (2.80), to the remarkable relation

$$\oint K d S = 2\pi\chi, \tag{9.5}$$

which relates the *total integrated Gaussian curvature* of any closed surface to the Euler characteristic of the surface. Thus, the total integrated Gaussian curvature of a closed surface is a topological invariant.

Chapter 10

Fitting of curves and surfaces

Very many problems of form description are particular cases of the following general problem: how can one construct a mapping, when the images of a given set of points are given? That is, a mapping is to be designed to fit a given incomplete description. Of course, there is no unique solution to such a problem; the characteristics of the chosen solution are determined by the particular problem context.

The simplest example is the problem of *interpolation* for a function of a single variable (mapping from a one-dimensional space to another). The values f_i of an unknown function, at a set of values x_i, are given, and an estimate of the values of the function at intermediate values of x is required.

The need to construct a smooth surface, with a given boundary and passing through a given set of points in space, is a problem encountered in the computer-aided design of aircraft shapes, ship hulls, car bodies, etc. The methods used are generalisations of the idea of 'interpolation' to mappings from two dimensions to three.

10.1 Interpolation

The simplest problem of interpolation is the problem of constructing a function $y = f(x)$ which will have the given values f_i at a set of given points x_i ($i = 1, \ldots n$) (the *sample* points). An obvious solution is *linear interpolation* [C5]; the curve $y = f(x)$ is taken to be a polygonal arc with the points (x_i, f_i) as its vertices. Explicitly, in the interval $[x_i, x_{i+1}]$,

$$f(x) = \alpha f_i + \beta f_{i+1} \tag{10.1}$$

$$\left. \begin{aligned} \alpha &= h_i^{-1}(x_{i+1} - x), \\ \beta &= h_i^{-1}(x - x_i) \end{aligned} \right\} \tag{10.2}$$

$$h_i = x_{i+1} - x_i \tag{10.3}$$

Alternatively, we can write, for the whole interval $[x_1, x_n]$,

$$f(x) = \sum_{i=1}^{n} W^i f_i \tag{10.4}$$

where the *basis functions* $W^i(x)$ are equal to 1 at x_i and zero at the remaining $n - 1$ given points, and vanish identically outside $[x_{i-1}, x_{i+1}]$ (where $x_0 = x_1$, $x_{n+1} = x_n$ by definition). W^1 and W^4 for example are shown in Fig. 181.

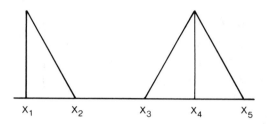

Figure 181

If the purpose of interpolation is to find an approximation to an unknown curve that passes through the n points in a 'well-behaved' way, then linear interpolation is not very accurate. The estimate of the unknown function at any point takes account only of the two neighbouring data points. A better approximation is obtained by drawing, by eye, a smooth curve through the given points. Mathematical methods of computing such a smooth curve through a given set of points are based on the fact that a polynomial of degree $n - 1$ has n coefficients, and therefore *there is a unique polynomial $f(x)$ of degree $n - 1$ through n given points (x_i, f_i).* Specifically,

$$\left.\begin{array}{l} f(x) = \sum_{i=1}^{n} f_i p_i(x) \\[2mm] p_i(x) = \prod_{j \neq 1} \dfrac{x - x_j}{x_i - x_j} \end{array}\right\} \tag{10.5}$$

Equations (10.1, 2, 3) refer to the particular case $n = 2$. Note that the polynomial p_i is equal to 1 at x_i and is zero at the other data points:

$$p_i(x_j) = \delta_{ij} \tag{10.6}$$

If the derivatives g_i, as well as the actual values f_i, of an unknown function are known at n points x_i, an approximation that has the correct values and the correct slopes at the data points has the form

$$f(x) = f_i \alpha + f_{i+1} \beta + g_i \gamma + g_{i+1} \delta \tag{10.7}$$

on the interval $[x_i, x_{i+1}]$ where the functions $\alpha, \beta, \gamma, \delta$ and their derivatives satisfy

$$\left.\begin{array}{llll}
\alpha(x_i) = 1, & \alpha(x_{i+1}) = 0, & \alpha'(x_i) = 0, & \alpha'(x_{i+1}) = 0, \\
\beta(x_i) = 0, & \beta(x_{i+1}) = 1, & \beta'(x_i) = 0, & \beta'(x_{i+1}) = 0, \\
\gamma(x_i) = 0, & \gamma(x_{i+1}) = 0, & \gamma'(x_i) = 1, & \gamma'(x_{i+1}) = 0, \\
\delta(x_i) = 0, & \delta(x_{i+1}) = 0, & \delta'(x_i) = 0, & \delta'(x_{i+1}) = 1.
\end{array}\right\} \quad (10.8)$$

If they are *cubics*, the solution is unique, and is given by

$$\left.\begin{array}{l}
\alpha = h_i^{-3} (x_{i+1} - x)^2 (h + 2(x - x_i)) \\
\beta = h_i^{-3} (x - x_i)^2 (h + 2(x_{i+1} - x_i)) \\
\gamma = h_i^{-3} (x - x_i)(x_{i+1} - x)^2 \\
\delta = h_i^{-3} (x - x_i)^2 (x - x_{i+1}).
\end{array}\right\} \quad (10.9)$$

(see Fig. 182).

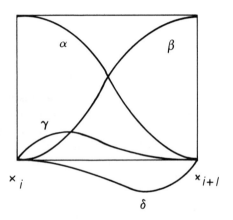

Figure 182

When n is large, the construction of the polynomial (10.5) of order $n - 1$ becomes excessively cumbersome, and a compromise is sought. Linear interpolation gives a curve that is continuous but not differentiable. A prescription of the form (10.1), but with α and β *quadratic* rather than linear, can be chosen so that the derivative of $f(x)$ is continuous; that is, the slopes match across the data points. With quadratic α and β, the requirement that $f(x)$ be equal to f_i at the points x_i implies the form

$$f(x) = h_i^{-1} (f_i(x_{i+1} - x) + f_{i+1}(x - x_i)) + c_i(x_{i+1} - x)(x - x_i), \quad (10.10)$$

on the interval $[x_i, x_{i+1}]$. The requirement that the slopes shall agree at the $n - 2$ internal data points leads to $n - 2$ equations for the $n - 1$ unknown constants c_i. In the case of equal intervals, these equations are

$$c_i + c_{i-1} = h^{-2} (f_{i+1} - 2f_i + f_{i-1}) (i = 2, \ldots n - 1). \quad (10.11)$$

One additional piece of information (for example, a specification of the second derivative at one point) will render the solution unique. A piecewise polynomial curve, with derivatives up to a specified order required to be continuous across the junctions, is a *spline* [A1]. The construction of splines is the most common method of interpolation for functions of a single variable. The most popular method makes use of *cubic splines*, whereby a curve is approximated by a cubic curve on each interval, and the first *and second* derivatives are made to match at each junction. Two additional pieces of information are required to make the solution unique; this is usually achieved by requiring the second derivative to vanish at the end points (x_i and x_n).

10.2 Bivariate Interpolation

Since a unique plane exists through any three points in 3-space, it follows that a function $f(x, y)$ of two variables, whose values f_i are known at n given points (x_i, y_i), can be interpolated as follows. Construct a set of triangles in the (x, y)-plane, whose vertices are the n given points. Then on each triangular region construct the unique linear function that takes the values f_i at the three vertices (Fig. 183). A different triangulation based on the same set of data points (x_i, y_i) will of course give a different polygonal surface. A computer method of obtaining a triangulation with a given set of points as vertices has been devised by Düppe and Gottschalk [D9]. The lengths of all lines joining pairs of the points

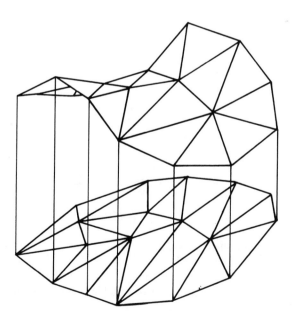

Figure 183

are computed, and ordered according to length. The shortest edge becomes a triangle side. The shortest remaining edge having a vertex in common with the first side becomes a second triangle side, and the first triangle is identified. All edges which intersect its sides are then discarded, and the process is repeated. The resulting triangulation was shown to be the one with minimum sum of edge lengths.

The linear mapping

$$
\left.\begin{aligned}
x &= (x_1 - x_3)u + (x_2 - x_3)v + x_3 \\
y &= (y_1 - y_3)u + (y_2 - y_3)v + y_3
\end{aligned}\right\} \tag{10.12}
$$

between an (x, y)-plane and a (u, v)-plane will map the general triangle with vertices (x_i, y_i) $(i = 1, 2, 3)$ to the triangle in the (u, v)-plane with vertices (10) (01) and (00) (which we denote by P_1, P_2 and P_3, respectively — Fig. 184). The linear function $f(u, v)$ that takes the values f_i at P_i $(i = 1, 2, 3)$ is then

$$
f(u, v) = \sum_{i=1}^{3} f_i p_i \tag{10.13}
$$

where

$$
\left.\begin{aligned}
p_1 &= u \\
p_2 &= v \\
p_3 &= u + v - 1
\end{aligned}\right\} \tag{10.14}
$$

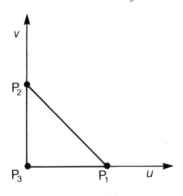

Figure 184

This gives the required linear interpolation for the interior of the given triangle, in terms of the new coordinates (u, v). If the values of $f(x, y)$ are given at the vertices and mid-points of the sides of a triangle, we can obtain a quadratic interpolation (a quadratic function of two variables has six coefficients):

$$
f(x) = \sum_{i=1}^{6} f_i Q_i \tag{10.15}
$$

where

$$
\left.\begin{aligned}
Q_1 &= u(2u - 1) = p_1(2p_1 - 1) \\
Q_2 &= v(2v - 1) = p_2(2p_2 - 1) \\
Q_3 &= (u + v - 1)(2u + 2v - 1) = p_3(2p_3 + 1) \\
Q_4 &= 4v(u + v - 1) = 4p_2 p_3 \\
Q_5 &= 4u(u + v - 1) = 4p_1 p_3 \\
Q_6 &= 4uv = 4p_1 p_2
\end{aligned}\right\} \qquad (10.16)
$$

(see Fig. 185). The factors of the quadratic basis functions are easily deduced from the observation that each Q_i must be equal to 1 at P_i and equal to zero at the other five points, which always lie on a pair of lines. The continuity of a quadratic interpolation of this kind, across the boundaries between triangular regions, is ensured because the function $f(x, y)$ so constructed reduces to a quadratic function of one variable on each triangle side, and there is a unique quadratic function of one variable that interpolates three data points.

Similarly, since a cubic polynomial in two variables has ten coefficients, a piecewise cubic interpolation can be constructed through ten values located at vertices, points of trisection of the sides, and the centroid, of a triangle in the (x, y)-plane. The image in the (u, v)-plane for a general triangle is shown in Fig. 186. The generalisation to higher orders is now obvious. These methods of interpolation for triangular regions, and their obvious extensions to tetrahedral regions in three dimensions, are employed in the 'finite element method' of numerical solution of partial differential equations [M10].

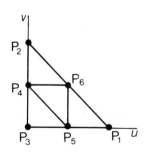

Figure 185

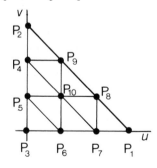

Figure 186

When the data points to be interpolated are arranged in a rectangular lattice, methods of interpolation that proceed from a triangulation are no longer appropriate. Fig. 187 illustrates a set of values f_{ij} located at the points (x_i, y_j) of a rectangular lattice, with intermediate values located on the edges of the rectangles interpolated linearly. This is the 'fishnet' representation of a surface $z = f(x, y)$, frequently employed in computer graphics. The interpolation can be completed

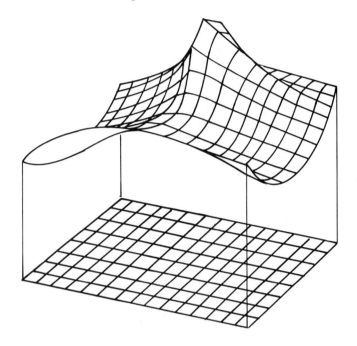

Figure 187

by constructing a quadratic function over each rectangular region. More generally, consider a tessellation of the (x,y)-plane into arbitrary convex quadrilaterals, with the values of an unknown function given at the vertices. Let $f_i(i = 1, \ldots 4)$ be the values at the vertices $P_i(x_i, y_i)$ $(i = 1, \ldots 4)$ of one quadrilateral. The *biquadratic transformation*

$$\left. \begin{aligned} x &= uvx_1 + (1-u)vx_2 + (1-u)(1-v)x_3 + u(1-v)x_4 \\ y &= uvy_1 + (1-u)vy_2 + (1-u)(1-v)y_3 + u(1-v)y_4 \end{aligned} \right\} \quad (10.17)$$

will map the quadrilateral to the square with vertices $(11), (01), (00), (10)$ in the (u, v)-plane (Fig. 188). Then

$$f = \sum_{i=1}^{4} f_i \varphi_i, \quad (10.18)$$

where

$$\left. \begin{aligned} \varphi_1 &= uv \\ \varphi_2 &= (1-u)v \\ \varphi_3 &= (1-u)(1-v) \\ \varphi_4 &= u(1-v), \end{aligned} \right\} \quad (10.19)$$

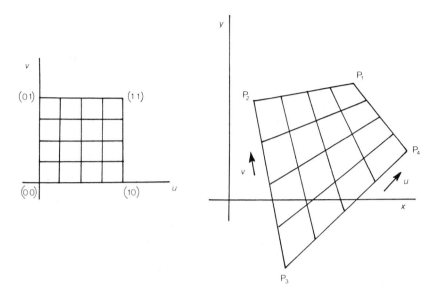

Figure 188

provides an interpolation, passing through the four values f_i at the points P_i, that is *linear* along the edges of the quadrilateral. Hence the piecewise interpolation will be continuous across the edges of the quadrilateral regions. The surface patches are hyperbolic paraboloids. The interpolated surface is of the same general type as the one illustrated in Fig. 174.

The mapping (10.17) is not the only way of mapping a quadrilateral in the (x, y)-plane to the standard square in the (u, v)-plane. An obvious alternative would be to use a *projective* mapping for this purpose (Fig. 189). The interpolated patches are still hyperbolic paraboloids, but of a slightly different shape.

Quadratic interpolation of the kind described here is the simplest way of constructing a surface $z = f(x, y)$ whose values at the vertices of a rectangular lattice are given. The resulting surface is continuous but not differentiable at the lattice edges. The construction of a continuous *differentiable* interpolated surface can be carried out by the more sophisticated methods of surface splines [A1].

10.3 Coons' Surface Patches

The computer-aided design and computer graphics display of curved surfaces usually rely on the methods developed by Coons [C9, C10], or on some variant of Coons' techniques.

A closed curve in 3-space, without self-intersections, can be mapped continuously onto the perimeter of the unit square (vertices (00), (01), (11), (10)) in a

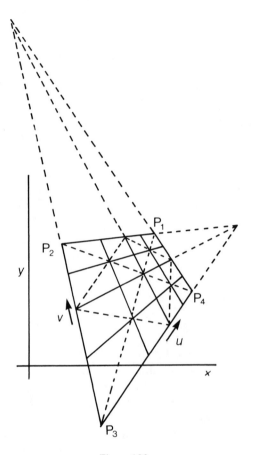

Figure 189

(u, v)-plane. This yields a description of the closed curve, which consists of four pieces

$$\left. \begin{array}{l} \mathbf{r} = \mathbf{P}(u, 0) \\ \mathbf{r} = \mathbf{P}(u, 1) \\ \mathbf{r} = \mathbf{P}(0, v) \\ \mathbf{r} = \mathbf{P}(1, v) \end{array} \right\} \tag{10.20}$$

$(0 < u < 1, 0 < v < 1)$. The four vertices of the square correspond to the four points $\mathbf{P}(00)$, $\mathbf{P}(01)$, $\mathbf{P}(11)$, $\mathbf{P}(10)$ on the curve. As is easily verified, the surface

$$\mathbf{Q}(u, v) = \mathbf{P}(u, 0)(1 - v) + \mathbf{P}(u, 1)v + \mathbf{P}(0, v)(1 - u) + \mathbf{P}(1, v)u$$
$$- \mathbf{P}(00)(1 - u)(1 - v) - \mathbf{P}(01)(1 - u)v - \mathbf{P}(10)u(1 - v) - \mathbf{P}(1, 1)uv$$

$$\tag{10.21}$$

contains the given curve. In matrix notation,

$$Q(u,v) = (1-u,u,1) \begin{pmatrix} -P(0\,0) & -P(0\,1) & P(0\,v) \\ -P(1\,0) & -P(1\,1) & P(1\,v) \\ P(u\,0) & P(u\,1) & 0 \end{pmatrix} \begin{pmatrix} 1-v \\ v \\ 1 \end{pmatrix} \quad (10.22)$$

This prescription for a surface patch with a given boundary is the simplest example of a Coons surface patch. The boundary consists of the four coordinate curves $v = 0, v = 1, u = 0$ and $u = 1$, which are the four curves of equation (10.20) (Fig. 190). The prescription is not unique. A different parametrisation of the four edges (satisfying the requirement that the parameters shall run from 0 to 1) will produce a different surface with the same boundary. A standardised patch with a given set of four space curves as boundary is obtained by requiring the

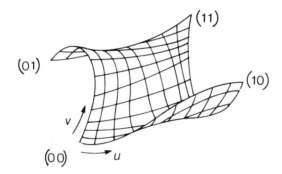

Figure 190

parameter on each of the four boundary curves to be s/L, where s is edge-length measured from an end-point and L is the total length of the curve. This method of obtaining a standardised patch is the method of *proportional development*. If the four boundary curves are straight lines, it produces a hyperbolic paraboloid, and equation (10.22) reduces to

$$Q(u,v) = (1-u,u) \begin{pmatrix} P(00) & P(01) \\ P(10) & P(11) \end{pmatrix} \begin{pmatrix} 1-v \\ v \end{pmatrix} = (u,1)\,MBM^T(v,1). \quad (10.23)$$

where

$$M = \begin{pmatrix} -1 & 1 \\ 1 & 0 \end{pmatrix} \quad (10.24)$$

and

$$B = \begin{pmatrix} P(00) & P(01) \\ P(10) & P(11) \end{pmatrix} \quad (10.25)$$

By regarding the vectors in (10.21) as vectors in a two-dimensional space, proportional development specifies a standard mapping of the interior of any 'curvilinear quadrilateral' in the (x, y)-plane onto the interior of a unit square in the (u, v)-plane (Fig. 191).

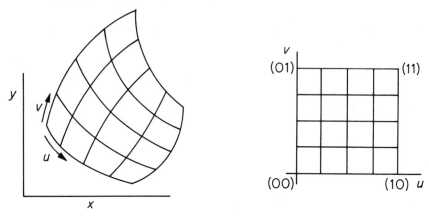

Figure 191

Fig. 192 shows the front-view of a proportionally-developed surface based on the boundary curves of a mould for the bow of a small boat (it is taken from Duncan and Vickers [D8]). This provides a *first approximation* to the shape of the mould. The actual surface to be designed is required to pass as closely as possible through a set of specified points in 3-space, as well as passing through the boundary curves. This is achieved by modifying the proportionally-developed surface $Q(u, v)$ by adding to the z-component of each of its points a bi-β function

$$A u^{\lambda_1} (1 - u)^{\lambda_2} v^{\lambda_3} (1 - v)^{\lambda_4} \qquad (10.26)$$

The general appearance of these functions is indicated in Fig. 193. Since the bi-β function vanishes on the boundary of the unit square, the boundary curves of the proportionally-developed surface will not be altered by the modification. The parameters of the bi-β function are adjusted so that the modified surface passes as closely as possible through the pre-assigned spatial points.

The fact that a different surface patch with the same boundary can be produced by changing the parametrisation of the boundary curves is equivalent to the fact that the prescription (10.22) can be generalised to

$$Q(u, v) = (\beta_0(u), \beta_1(u), 1) \begin{pmatrix} -P(00) & -P(01) & P(0v) \\ -P(10) & -P(11) & P(1v) \\ P(u0) & P(u1) & 0 \end{pmatrix} \begin{pmatrix} \gamma_0(v) \\ \gamma_1(v) \\ 1 \end{pmatrix} \qquad (10.27)$$

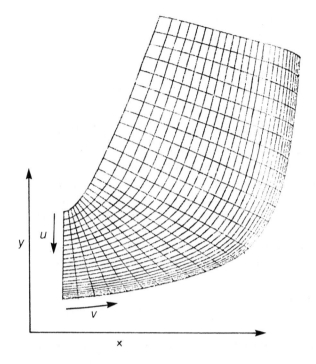

Figure 192

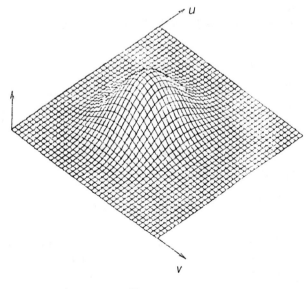

Figure 193

where the *blending functions* β_0, β_1, γ_0, γ_1 have no maximum or minimum in the interval $[0, 1]$ and satisfy

$$\left. \begin{aligned} \beta_0(0) = \beta_1(1) = \gamma_0(0) = \gamma_1(1) = 1 \\ \beta_0(1) = \beta_1(0) = \gamma_0(1) = \gamma_1(0) = 0 \end{aligned} \right\} \tag{10.28}$$

and

$$\beta_0(u) + \beta_1(u) = \gamma_0(v) + \gamma_1(v) = 1 \tag{10.29}$$

The equation (10.29) is necessary in order that the surface patch prescribed by (10.27) shall be independent of the choice of origin in 3-space, and also ensures that the surface patch will be planar if its boundary lies in a plane.

Coons surface patches can be fitted together along boundary curves to produce continuous surfaces containing a given reticulated net of space-curves. Such a surface will not, of course, be *differentiable* across the boundary curves between patches. Generalisations of the methods are concerned with means of overcoming this lack of differentiability. For example, consider the problem of defining a surface patch when all the quantities

$$\left(\frac{\partial}{\partial u} \right)^i \left(\frac{\partial}{\partial v} \right)^j \mathbf{Q}(u, v) \quad (i, j = 0 \text{ or } 1) \tag{10.30}$$

are specified *on the boundaries*. Let their specified values be denoted by

$$\mathbf{P}^{ij}(u, k), \quad \mathbf{P}^{ij}(k, v) \quad (k = 0, 1). \tag{10.31}$$

Then

$$\mathbf{Q}(u, v) = [\beta_{00}(u)\beta_{01}(u)\beta_{10}(u)\beta_{11}(u)1] \, \mathbf{A} \, [\gamma_{00}(v)\gamma_{01}(v)\gamma_{10}(v)\gamma_{11}(v)1]^{\mathrm{T}} \tag{10.32}$$

provides a solution, where $\mathbf{A}$ is the matrix

$$\mathbf{A} = \begin{pmatrix} -\mathbf{P}(00) & -\mathbf{P}(01) & -\mathbf{P}^{01}(00) & -\mathbf{P}^{01}(01) & \mathbf{P}(0v) \\ -\mathbf{P}(10) & -\mathbf{P}(11) & -\mathbf{P}^{01}(10) & -\mathbf{P}^{01}(11) & \mathbf{P}(1v) \\ -\mathbf{P}^{10}(00) & -\mathbf{P}^{10}(01) & -\mathbf{P}^{11}(00) & -\mathbf{P}^{11}(01) & \mathbf{P}^{10}(0v) \\ -\mathbf{P}^{10}(10) & -\mathbf{P}^{10}(11) & -\mathbf{P}^{11}(10) & -\mathbf{P}^{11}(11) & \mathbf{P}^{10}(1v) \\ \mathbf{P}(u0) & \mathbf{P}(u1) & \mathbf{P}^{01}(u0) & \mathbf{P}^{01}(u1) & 0 \end{pmatrix} \tag{10.33}$$

and the blending functions $\beta_{ij}(u)$ and $\gamma_{ij}(v)$ satisfy

$$\left(\frac{\partial}{\partial u} \right)^k \beta_{ij}(l) = \left(\frac{\partial}{\partial v} \right)^k \gamma_{ij}(l) = \delta_{ik} \delta_{jl} \quad (i, j, k, l = 0, 1). \tag{10.34}$$

As a second example, suppose the quantities (10.30) are given at just four points corresponding to the values (00), (01), (11) and (10) of (u, v), without specification of boundary curves. Then

$$\mathbf{P}(u, v) = [u^3 u^2 u \, 1] \, \mathbf{MBM}^{\mathrm{T}} [v^3 v^2 v \, 1]^{\mathrm{T}} \tag{10.35}$$

is a solution, where

$$M = \begin{pmatrix} 2 & -2 & 1 & 1 \\ -3 & 3 & -2 & -1 \\ 0 & 0 & 1 & 0 \\ 1 & 0 & 0 & 0 \end{pmatrix} \qquad (10.36)$$

and

$$B = \begin{pmatrix} P(00) & P(01) & P^{01}(00) & P^{01}(01) \\ P(10) & P(11) & P^{01}(10) & P^{01}(11) \\ P^{10}(00) & P^{10}(01) & P^{11}(00) & P^{11}(01) \\ P^{10}(10) & P^{10}(11) & P^{11}(10) & P^{11}(11) \end{pmatrix} \qquad (10.37)$$

This is the *bicubic patch*, which is very commonly used in computer-aided design. A pair of bicubic patches with a common pair of vertex points will join in a continuous and differentiable manner along a common boundary curve.

For further details of the methods of Coons patches, the reader is referred to Forrest [F3]. This work contains an extensive bibliography of methods and applications in surface description.

10.4 Approximate Methods of Curve Fitting

The previous sections of this chapter have dealt with a set of problems concerned with the construction of a mapping to fit a given set of data exactly. A different set of problems is concerned with constructing a mapping, of a pre-determined type, which will approximately fit a given set of data 'as closely as possible'.

The simplest example of such a problem is the determination of the 'best straight line' through a given set of values f_i of an unknown function $y(x)$ at the points x_i. For example, the f_i may be experimentally determined values of a variable $y(x)$ which is expected to be linearly (or approximately linearly) dependent on x. Since the f_i are known to contain errors, a curve passing exactly through the points (x_i, f_i) is not appropriate.

A popular method of approach to the above problem is the *method of least squares*. Let

$$y = \phi(x) = \alpha + \beta x \qquad (10.38)$$

be the equation of the 'best' straight line approximation to the data points. The unknown constants α and β are determined so as to minimise the sum of squares of the deviations of $\phi(x_i)$ from the f_i. That is, the function

$$\chi = \sum_i (\phi(x_i) - f_i)^2, \qquad (10.39)$$

regarded as a function of α and β, is to be a minimum (Fig. 194).

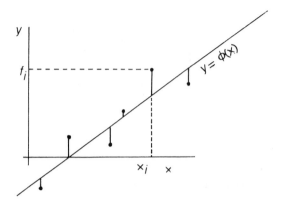

Figure 194

The conditions

$$\frac{1}{2}\frac{\partial \chi}{\partial \alpha} = \sum_i (\phi(x_i) - f_i) = 0$$

$$\frac{1}{2}\frac{\partial \chi}{\partial \beta} = \sum_i (\phi(x_i) - f_i)x_i = 0$$

(10.40)

are a pair of simultaneous linear equations for the unknowns α and β. The solution is

$$\left.\begin{array}{l} \alpha = \begin{vmatrix} n & \Sigma x_i \\ \Sigma x_i & \Sigma x_i^2 \end{vmatrix}^{-1} \cdot \begin{vmatrix} \Sigma f_i & \Sigma x_i \\ \Sigma f_i x_i & \Sigma x_i^2 \end{vmatrix} \\[3mm] \beta = \begin{vmatrix} n & \Sigma x_i \\ \Sigma x_i & \Sigma x_i^2 \end{vmatrix}^{-1} \cdot \begin{vmatrix} n & \Sigma f_i \\ \Sigma x_i & \Sigma f_i x_i \end{vmatrix}, \end{array}\right\}$$

(10.41)

where n is the number of data points.

The best *quadratic* approximation

$$\phi(x) = \alpha + \beta x + \gamma x^2$$

(10.42)

to the same data can be found by an obvious generalisation of the method, and similarly for higher order polynomials. If $\phi(x)$ is chosen to be a polynomial of degree $n - 1$, the method will, of course, yield the solution (10.5) which passes precisely through the n data points. The method can be modified by multiplying the terms in (10.39) by weights w_i, so as to assign different degrees of importance to the data points. This would be appropriate, for example, if the errors in the data points are different.

A different kind of least squares approximation is appropriate if the 'best straight line' through a set of points (x_i, y_i) in a Euclidean plane is required. The quantity to be minimised is then the sum of the squares of the perpendicular

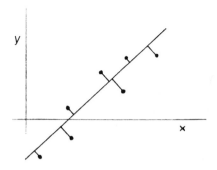

Figure 195

distances of data points from the line (Fig. 195), not the sum of squares of deviations of their y-coordinates. (This method is appropriate when there are errors in both the f_i and x_i, and the expected values of these errors are known. The axes must first be normalised so that the expected values of the errors in the f_i and the x_i are equal.) The required line passes through the mean position of the data points, and its normal is the eigenvector associated with the smallest eigenvalue of the matrix

$$\begin{pmatrix} \Sigma x_i^2 & \Sigma x_i y_i \\ \Sigma x_i y_i & \Sigma y_i^2 \end{pmatrix} \tag{10.43}$$

The fitting of a high order polynomial to a large number of data points, by minimising (10.39), is excessively cumbersome. The need for efficient computational methods then becomes an important criterion in choosing appropriate methods of approximation. A piecewise method of fitting a continuous curve approximately to a set of data points (x_i, f_i), analogous to the method of splines for exact fitting, is as follows. Let the x-axis be divided into m intervals, with boundaries x_α ($\alpha = 1 \ldots m$), so that there are several data points in each region. Then in any pair of consecutive intervals, $[x_{\alpha-1}, x_{\alpha+1}]$, one can fit a continuous function $\phi_\alpha(x)$ (for example, by the least squares polynomial method). Then, on any interval $[x_\alpha, x_{\alpha+1}]$, with the exception of the two end intervals, there are *two alternative* approximating functions $\phi_{\alpha-1}(x)$ and $\phi_\alpha(x)$. The approximation to the data in the interval is taken to be a weighted mean

$$\psi_\alpha(x) = \phi_{\alpha-1}(x)\, w_1(x) + \phi_\alpha(x)\, w_2(x). \tag{10.44}$$

If the weighting functions are linear and we require that the pieces shall join up to give a continuous curve $(f_{\alpha-1}(x_\alpha) = f_\alpha(x_\alpha))$, then

$$\left. \begin{array}{l} w_1 = 1 - \xi \\ w_2 = \xi \end{array} \right\} \quad \xi = \frac{x - x_\alpha}{x_{\alpha+1} - x_\alpha} \tag{10.45}$$

(see Fig. 196a, b). If the final approximation is required to have a continuous first derivative as well as being continuous, then the weighting functions have to be cubics.

$$w_1 = \xi^2 (3 - 2\xi) \quad \Big\}$$
$$w_2 = 1 - \xi^2 (3 - 2\xi) \Big\} \qquad (10.46)$$

(Fig. 196c). This method of approximate curve fitting by weighted averages of overlapping preliminary approximations extends readily to higher dimensions, and has been successfully exploited by Junkins, Miller and Jancaitis [J4, P6] as a means of computer storage and manipulation of topographic information. Their method will be discussed in §10.7.

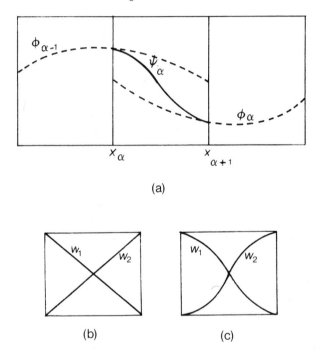

(a)

(b)   (c)

Figure 196

Sometimes the need for fast and efficient computational methods becomes a dominant restriction on a form description method. This is the case in many areas of pattern recognition by computers [P6, U1], when the amount of initial data is very large. Thus extremely simple and unsophisticated methods of curve description have been developed and found to be effective. We give one example, which was used by Groner [G7] in the context of a program for computer recognition of handwritten characters. The problem is to approximate to a sequence of points (x_i, y_i) by a polygonal arc which is 'smoother' than that obtained by linear interpolation. Labelling the points of the smoothed data (ξ_i, η_i), Groner's algorithm is simply

$$\xi_1 = x_1, \quad \eta_1 = y_1, \quad \xi_i = \tfrac{1}{4}(x_i + 3\xi_{i-1}), \quad \mu_i = \tfrac{1}{4}(y_i + 3\eta_{i-1}) \quad (10.47)$$

The kind of result obtained is indicated in Fig. 197.

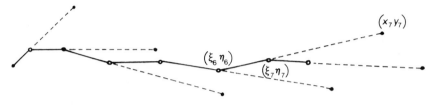

Figure 197

10.5 Smoothing

Closely related to the problem of approximating a curve to a set of data points (x_i, f_i), is the problem of *smoothing* a complicated function $y = f(x)$. For example, the 'best' polynomial approximation to a given function, by a polynomial of given degree (Fig. 198) can be obtained by the least-squares method already discussed, simply by changing the summation in (10.39) to integration over an interval of the x-axis. For example, the straight line that approximates the curve $y = f(x)$ over the interval $[0, 1]$, given by this method, is easily found to be $y = \alpha + \beta x$, where

$$\left.\begin{aligned}
\alpha &= 2 \int_0^1 f(\xi)\,(2 - 3\xi)\,d\xi, \\
\beta &= 6 \int_0^1 f(\xi)\,(2\xi - 1)\,d\xi.
\end{aligned}\right\} \qquad (10.48)$$

A commonly employed smoothing process is the computation of a *moving average*. This produces, from a given complicated function $f(x)$, a less erratic function $\tilde{f}(x)$ (whose maxima and minima will be flatter). The process is defined by

$$\tilde{f}(x) = \frac{1}{W} \int_{x - \frac{W}{2}}^{x + \frac{W}{2}} f(\xi)\,d\xi \qquad (10.49)$$

That is, the value of the function at any point x is replaced by the average of the function over a small interval, of width W, centred at x. A larger width will produce a 'smoother' result. The procedure will convert a discontinuous function to a continuous one, and a function with a discontinuous slope will be converted to one with continuous slope (Fig. 199).

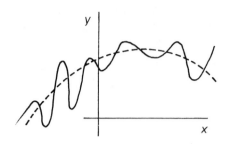

Figure 198

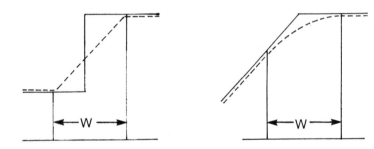

Figure 199

For a function of two variables $f(x, y)$, the value at each point of the (x, y)-plane is replaced by the average over a small square 'window' of side W, centre at each point (whose sides are parallel to the axes). If $f(x, y)$ is a 'picture function', describing the grey-levels of a photographic image, the effect of the process is something like that of using a camera out of focus. The purpose of the process is to eliminate irrelevant small scale random variations (noise). The optimum width is the one that achieves this without eliminating too much relevant information. Smoothing techniques more sophisticated than the one described here are based on Fourier techniques.

Geographical data is sometimes encountered in an aggregated form, the variable described being averaged over administrative areas, so that $f(x, y)$ is presented as a stepwise discontinuous function (Fig. 200a, b). A smoothing technique applied to such a stepped function can recover aspects of the data that have been obscured by the process of aggregation (Fig. 200c). Hsu and Robinson [H13, H14] have studied the nature of the distortion of information inherent in descriptions based on areal aggregation. Harmon and Julesz [H2, H3] have shown that the process of smoothing a digitised picture (e.g. Fig. 207) can render it more recognizable, thus indicating that the information loss due

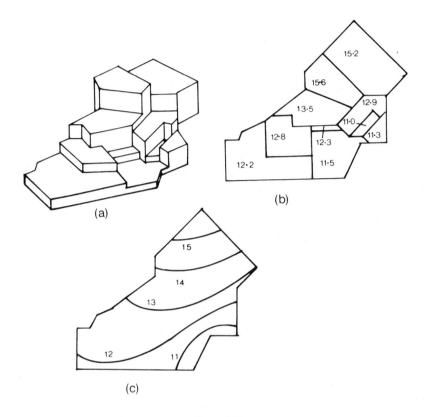

Figure 200

to digitisation or areal aggregation is not so great as might be supposed. The Shannon sampling theorem (described in §12.5) throws some light on this phenomenon.

10.6 Surface Modelling

Methods of fitting a continuous surface $z = \phi(x, y)$ to a set of data points (x_i, y_i, f_i), without requiring that it shall contain the points, are of great practical importance in computer-assisted cartography. A low-order polynomial least squares fit will give a surface which indicates the overall 'smoothed out' trend of the data, thus enabling this global information content of the data to be subtracted out, leaving only information concerned with smaller scale local variations and random errors. For example, a linear least-squares fit will specify a plane indicating the general slope characteristic of the surface under investigation. This plane is obtained in the form $z = \phi(x, y) = \alpha x + \beta y + \gamma$ by solving $\partial \chi / \partial \alpha = \partial \chi / \partial \beta = \partial \chi / \partial \gamma = 0$ for α, β and γ, where χ is analagous to (10.39). Polynomial surfaces can be fitted in an exactly analogous manner, but the number of simultaneous equations to be solved increases rapidly with the degree of the polynomial (a polynomial of degree N in two variables has $N!(N+2)!/2$ coefficients). For this reason, polynomials whose coefficients are not independent have been resorted to [G6, K9]. Surfaces other than polynomial surfaces have also been employed. The 'Fourier surfaces'

$$z = \sum_{k=-K}^{K} \sum_{l=-L}^{L} \alpha_{kl} \, e^{i(k\beta x + l\gamma y)} \quad (a_{kl} = -a^*_{-k\,-l}) \tag{10.50}$$

have been used as 'trend surfaces' to give least-squares fits to topographic data [J1]. The $KL + 1$ real parameters $a_{jk} - a_{-j-k}$ and a_{00} enter linearly like the coefficients of polynomial surfaces, but the parameters β and γ enter non-linearly in a complicated manner. Thus the method, to be effective, requires *a priori* specification of β and γ. The reader is referred to the review article of Whitten [W9] for further information about trend surface analysis.

    The above method of fitting a simple analytic surface with only a few free parameters to a relatively complicated numerically described surface yields only a very 'smoothed out' description of the broad features of the actual surface. This is, of course, the purpose of the method, the idea being to separate out these broad features from the small-scale local variations.

    If the purpose is to obtain, from a numerical description of a complicated surface, a continuous description that is more faithful to details, one has to resort to a piecewise surface fit, by surface splines or some similar technique. Dividing the sample plane (x, y) into regions and obtaining a least-squares polynomial approximation for each region will yield a set of approximating surfaces that do not fit together at the region boundaries. The method of Junkins et. al. [J4] overcomes this difficulty. The method proceeds by putting a rectangular grid over the sample plane, whose spacing is large enough so that any rectangular region consisting of four adjoining grid rectangles contains sufficient sample points to obtain a least-squares polynomial approximation.

    Label a block of nine grid rectangles as in Fig. 201, and denote the least squares polynomial approximations for the four regions 2356, 1245, 4578 and 5689 respectively by ϕ_1, ϕ_2, ϕ_3 and ϕ_4. Then we have *four alternative* approximations for the region 5. The final approximation for this region is taken to be a weighted mean

$$\psi_5(x, y) = W_1 \phi_1 + W_2 \phi_2 + W_3 \phi_3 + W_4 \phi_4. \tag{10.51}$$

The weighting functions are given in terms of (10.45) or (10.46) by

$$\left. \begin{array}{l} W_1(\xi, \eta) = w_2(\xi)\, w_1(\eta) \\ W_2(\xi, \eta) = w_1(\xi)\, w_2(\eta) \\ W_3(\xi, \eta) = w_1(\xi)\, w_1(\eta) \\ W_4(\xi, \eta) = w_2(\xi)\, w_2(\eta) \end{array} \right\} \tag{10.52}$$

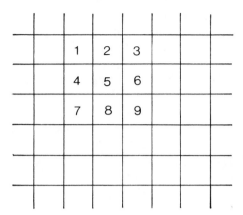

The quadratic functions obtained from the linear functions w_1 and w_2 are indicated by contours in Fig. 202. With quadratic weighting functions and linear preliminary approximations ϕ_1, ϕ_2, ϕ_3 and ϕ_4, the final approximation ψ_5 will be cubic. The quadratic weighting functions give a final approximation that is continuous at boundaries. The bicubic weighting functions give a final approximation that also has a continuous slope at boundaries. The surface description obtained can be stored in a computer either by storing the coefficients of the

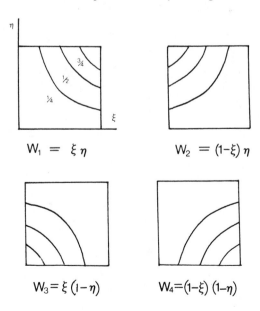

$$W_1 = \xi\eta \qquad\qquad W_2 = (1-\xi)\eta$$

$$W_3 = \xi(1-\eta) \qquad\qquad W_4 = (1-\xi)(1-\eta)$$

Figure 202

polynomials ψ, or by storing the coefficients of the preliminary polynomials ϕ and using the algorithm (10.51) when information about the surface is required. The latter method is superior in its efficient use of storage space, but requires more computation whenever information is needed.

The effect of changing the grid spacing is indicated in Fig. 203. A larger grid size has the effect of 'smoothing' the description, while a grid that is too fine for the density of sample points will misrepresent the data by producing spurious detailed variations. Thus there is an optimum grid size for any particular set of surface data. The smoothing obtained with larger grid sizes is a desirable feature in cartographic applications, where the problem is to reduce the scale of a contour map; of necessity a small-scale map cannot contain as much information as one at a larger scale, and one encounters the problem of *map generalisation* [T4]. The smoothing of contours is only one aspect of this problem,

Figure 203

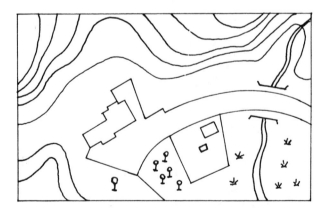

Figure 204

whose wider aspects include more general questions of simplifying cartographic information while retaining essential information content such as areas of objects appearing in the map (Fig. 204).

10.7 Statistical Methods

The form of certain phenomena (eg. turbulent fluid flow, 'noise' in a signal or picture function, etc.) is expressible only in terms of *random functions*. A random function is characterised by erratic variations which render complete description impossible, or at least irrelevant. Description of such a function consists of statistical information about the systematics of its behaviour, in terms of *expectation values* (i.e. averaged values over sufficiently large regions, over many instances of the same phenomenon, or over time). For example, let $f(x)$ be a random function and let E denote an expectation value. Then

$$E(f(x)) = m(x) \qquad (10.53)$$

is the *mean value* of the function at x,

$$K(x, y) = E(f(x)f(y) - m(x)m(y)) \qquad (10.54)$$

is the *covariance* between the values at two points, and $\gamma(h, x)$, defined by

$$2\gamma(h, x) = E[(f(x+h) - f(x))^2] \qquad (10.55)$$

is the *variogram*. The *variance*, or mean square deviation, of two random functions is

$$\sigma^2 = E[(f(x) - g(x))^2]. \qquad (10.56)$$

When faced with measurements $z_i = f(x_i, y_i)$ of a geographical phenomenon (e.g. rainfall, magnetic field, grade and thickness of ore, population density, etc.), it is often useful to be able to extract from the data estimates of statistical descriptors of structural features of the phenomenon, on the assumption that the underlying variable $f(x, y)$ is a random function [D2, D4, H16].

The theory of random functions leads to a method of surface fitting known as *Kriging* [D4, K8], which has been used with much success for geographical data of the kind mentioned above. For simplicity, we illustrate the basic method for the one-dimensional case. Let $z_i = f(x_i)$ be the measured data. Assume that the underlying function $f(x)$ is a random function. The idea is to estimate it by a random function $g(x)$ which has the *same mean* ('unbiased' estimation) and such that the variance is as small as possible. Assume that the mean is a polynomial of degree $n + 1$:

$$m(x) = \sum_{\alpha=0}^{n} a_\alpha x^\alpha \qquad (10.57)$$

and let the required estimator be of the form

$$g(x) = \sum_i W^i(x) f(x_i) \qquad (10.58)$$

The equality of means leads to

$$\sum_i W^i x_i^\alpha = x^\alpha. \qquad (10.59)$$

The functions W^i are chosen so as to minimise the variance (10.56), subject to the restriction (10.59). This leads to

$$\sum_j W^j K(x_i x_j) - \sum_\alpha \mu_\alpha x_i^\alpha = K(x_i x) \qquad (10.60)$$

(where the $\mu_\alpha(x)$ are Lagrange multipliers). We now have a set of simultaneous linear equations (10.59) and (10.60) for the unknowns W^i and μ_α. The minimum value of the variance is found to be

$$\sigma^2 = K(x, x) - \sum_i W^i K(x_i x) + \sum_\alpha \mu_\alpha x^\alpha. \qquad (10.61)$$

In order to find the required functions W^i, a knowledge of the covariances $K(x_i x_j)$ and $K(x_i x)$ is required. This requires further assumptions. We assume:

$$\left.\begin{array}{l} m(x) \text{ constant,} \\ K(x, y) = K(x - y), \\ \gamma(h, x) \text{ independent of } x. \end{array}\right\}$$

This leads to

$$K(x, y) = -\gamma(x - y). \qquad (10.62)$$

Thus, given an estimate of the function $\gamma(h)$ from the data (for example by a simple polynomial fit), the equations (10.59) and (10.60) can be solved for the W^i, and (10.61) will give an indication of the accuracy of the fit of the estimated curve $z = \sum_i W^i(x) z_i$ to the data.

One important advantage of the method is that it provides an indication, by means of the computed variance, of the error involved. That is, it gives an idea of the deviation of the estimated curve or surface from the true (unknown) curve or surface. The only indication of the inaccuracy of a polynomial method is the deviation of the estimator *from the given data*, which is in general an underestimation of the true error.

10.8 Linear Operators
There is a single unifying concept underlying the methods that have been discussed in this chapter. This is the idea of obtaining, from a given description, an alternative description, by means of a linear operator. For example, a discrete description consisting of a set of n descriptors $f_i(i = 1, \ldots n)$ can be transformed to a new discrete description consisting of a set of m descriptors $\overline{f}_\alpha(\alpha = 1, \ldots m)$, by matrix multiplication,

$$\overline{f}_\alpha = \sum_i W_\alpha^i f_i \tag{10.63}$$

If the rank of the matrix W is less than n, some information is lost in going to the new description; also, if $m > n$, the new description will contain some redundant information. A particular case is hat of a square, nonsingular matrix; the process is then invertible and the two descriptions are equivalent.

An example of information-processing of the type (10.63) is the construction of data about a surface $z = f(x, y)$ at a lattice of sample points, from an initial description based on random sample points (see §11.1).
If

$$\sum_i W_\alpha^i = 1, \tag{10.64}$$

each value $\widetilde{f}_\alpha$ is a *weighted mean* of the values f_i.

Interpolation, and estimation by curve or surface fitting, are generalizations in which the final description is a continuous description in terms of a function of one or more variables,

$$\overline{f}(x) = \sum_i W^i(x) f_i. \tag{10.65}$$

Alternatively, the initial description may be continuous and the final description discrete,

$$\overline{f}_\alpha = \int W_\alpha(x) f(x)\, dx \tag{10.66}$$

For example, let the interval of the x-axis over which $f(x)$ is defined be divided into m intervals by the points $x_\alpha(\alpha = 1, \ldots m + 1)$ and let the *weight functions* be

$$W_\alpha(x) = \begin{cases} 1/(x_{\alpha+1} - x_\alpha) \text{ if } x \epsilon [x_\alpha, x_{\alpha+1}] \\ \qquad 0 \qquad \text{otherwise} \end{cases} \tag{10.67}$$

Then (10.66) specifies the process of aggregation of data (Fig. 205). Areal aggregation (Fig. 200a, b) corresponds to

$$f_\alpha = \int_{R\alpha} W_\alpha(x, y) f(x, y)\, dA \tag{10.68}$$

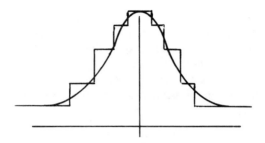

Figure 205

where the region of the (x, y) plane is tessellated into regions R_α, with areas A_α, and

$$W_\alpha(x, y) = \begin{cases} 1/A_\alpha & \text{if } (x, y) \in R_\alpha \\ 0 & \text{otherwise.} \end{cases} \tag{10.69}$$

The *Dirac delta function* $\delta(x)$ is defined to be zero everywhere except at the origin, where it is infinite, and

$$\int \delta(x)\, dx = 1 \tag{10.70}$$

when the interval of the integration contains the origin. Therefore it follows that, if the weight functions in (10.66) are defined to be

$$W_\alpha(x) = \delta(x - x_\alpha), \tag{10.71}$$

then

$$\overline{f}_\alpha = f(x_\alpha). \tag{10.72}$$

Thus the delta functions (10.71) provide a description of the sampling of information at a discrete number of points.

A particular example of a linear operation of the form (10.66) is the construction of the *moments* of a function. This corresponds to choosing the weighting functions to be powers of x,

$$W_\alpha(x) = x^\alpha \quad (\alpha = 0, 1, \dots). \tag{10.73}$$

In the case of a function $f(x, y)$ of two variables, the moments are

$$\mu_{\alpha\beta} = \int_{-\infty}^{\infty} \int_{-\infty}^{\infty} x^\alpha y^\beta f(x, y)\, dx dy. \tag{10.74}$$

The shape of a region of a Euclidean plane can be described by a *binary* function $f(x, y)$, which can take only two values, 1 and 0 according to whether the point (x, y) is in the region or not. In this case, μ_{00} is the *area* of the region and

(μ_{10}, μ_{01}) are the coordinates of its *centre of 'gravity'*. The application of a translation so that $\mu_{10} = 0$ has been used in character recognition problems to standardise the position of the character as a preliminary to the application of character-recognition techniques [H15, S12]. The moment of inertia of the shape about the origin is $\mu_{20} + \mu_{02}$. If the plane figure is the cross-section of a beam, the bending moment [N1] is

$$M = \iint y f(x, y) \, dxdy \qquad (16.75)$$

(the x-axis being mid-way between the upper and lower y-values). This is a particularly simple example of the extraction of information about a form, in order to obtain a form descriptor relevant to a particular problem context. Examples of the value of M for different shapes are given in Fig. 206.

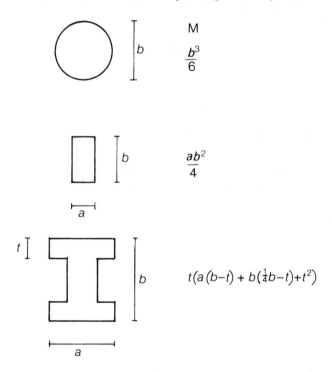

Figure 206

When the initial and the final descriptions are both continuous, a linear operator has the form

$$\overline{f}(u) = \int w(u, x) f(x) \, dx \qquad (10.76)$$

A special case of extreme importance is the *Fourier transform* of a function $f(x)$, which corresponds to $w(u, x) = e^{-iux}/2\pi$ and integration over the whole

x-axis. The *Fourier series* corresponds similarly to a special form of (10.65). Fourier methods will be dealt with in detail in chapter 12.

Another special form of (10.76) corresponds to the case where $w(u, x)$ has the special form $w(u, x) = w(u - x)$ and the integration is over the whole x axis. Then $\bar{f}$ is the *convolution* of the function $f(x)$ with the function $w(x)$, which is written $f * w$,

$$(f * w)(u) = \int_{-\infty}^{\infty} w(u - x) f(x) \, dx \qquad (10.77)$$

(The roles of the two functions are interchangeable, that is, $f * w = w * f$, as is easily established by a change of variable). A particular case of convolution is the smoothing operation by a *moving average* (10.49). This corresponds to the $\bar{f} = f * w$, with

$$w(x) = \begin{cases} 1/W \text{ if } x \in \left(-\dfrac{W}{2}, \dfrac{W}{2}\right), \\ 0 \quad \text{otherwise.} \end{cases} \qquad (10.78)$$

For functions of two variables, the convolution is

$$(f * w)(u, v) = \int_{-\infty}^{\infty} \int_{-\infty}^{\infty} w(u - x, v - y) f(x, y) \, dx \, dy, \qquad (10.79)$$

and the smoothing procedure by a moving average over a square 'window', described in §10.5, is given by $\bar{f} = f * w$, $w(x, y) = w(x) w(y)$, in terms of the functions (10.78). The more general cases of convolution can be regarded as the formation of a moving *weighted* average.

Chapter 11

The discretised euclidean plane

For purposes of numerical description, the space in which a form is situated has to be dealt with as a discrete space. The usual method of discretisation is the one exploited in the finite difference method of solving differential equations; the field governed by the equations is computed at a set of points whose coordinates are *integers*. Fig. 207, for example, indicates the appearance of such a *generalised*

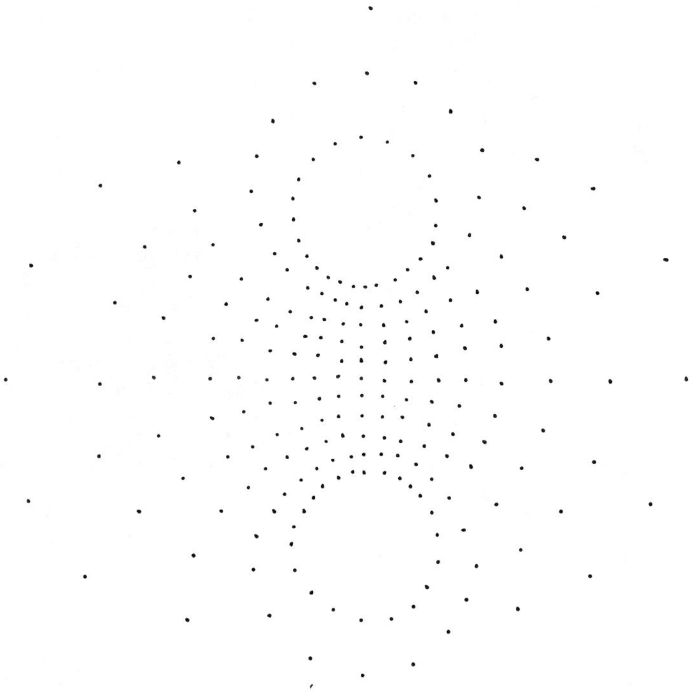

Figure 207

lattice, for the curvilinear coordinate system of Fig. 45. The points with integer coordinates in a Cartesian coordinate system form a regular square lattice (Fig. 140d). This chapter is concerned with the description of form on a regular square lattice.

11.1 Scalar Functions on a Square Lattice

The set of points P_{ij} in a Euclidean plane, whose Cartesian coordinates have the form (x_i, y_i), form a rectangular lattice. If the coordinates are restricted to *integer* values, we obtain a regular square lattice, whose points P_{ij} have coordinates (i, j).

A scalar field $f(x, y)$ on the plane can be approximately described by an array of numbers z_{ij}, in either of two ways:
(i) z_{ij} is the value of f at the point P_{ij}:

$$z_{ij} = f(i, j), \tag{11.1}$$

or
(ii) z_{ij} is the average value of f, over a unit square centred at P_{ij}:

$$z_{ij} = \int_{j-\frac{1}{2}}^{j+\frac{1}{2}} \int_{i-\frac{1}{2}}^{i+\frac{1}{2}} f(x, y)dxdy. \tag{11.2}$$

A typical example of the second kind of representation is the result of scanning a photograph for purpose of image-processing by computer. Here, $f(x, y)$ is a 'picture function' describing the grey-levels of the image [R5]. After 'rounding' the numbers z_{ij} (11.2) to a small number of values (e.g. the integers 0 to 10), the result is an array of integers (Fig. 208a). In Fig. 208b the same information has been presented by representing the integer values by the areas of small circles.

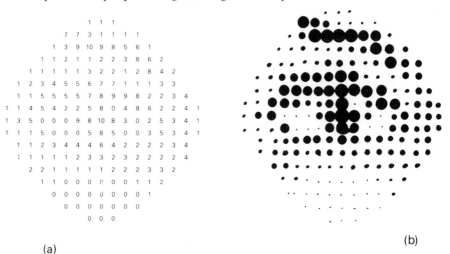

```
                1   1   1
             7  7   3   1   1   1   1
          1  3  9  10   9   8   5   6   1
          1  1  2   1   1   2   2   3   8   6   2
       1  1  1  1   1   3   2   2   1   2   8   4   2
       1  2  3  4   5   5   6   7   7   1   1   1   3   3
       1  1  5  5   5   5   7   8   9   9   8   2   2   3   4
    1  1  4  5  4   2   2   5   8   0   4   8   6   2   2   4   1
    1  3  5  0  0   0   9   8  10   8   3   0   2   5   3   4   1
    1  1  1  5  0   0   0   5   8   5   0   0   3   5   3   4   1
       1  1  2  3   4   4   4   6   4   2   2   2   2   3   4
       1  1  1  1   1   2   3   3   2   3   2   2   2   2   4
          2  2  1   1   1   1   1   2   2   2   3   3   2
             1  1   0   0   0   0   0   0   1   1   2
                0   0   0   0   0   0   0   0   1
                   0   0   0   0   0   0   0
                      0   0   0
```

(a)   (b)

Figure 208

When the values of a function $f(x, y)$ are known at a set of random sample points, a common requirement is the conversion of this information to a (necessarily approximate) specification of the values of the function at the points of a regular square lattice [D6, R5]. This would be so, for example, if the preliminary data were the result of a land survey, where height above sea level would be known at a 'random' set of sample points (random only in the sense that they do not form a regular lattice. Their position is chosen according to a visual assessment of the nature of the terrain; the surveyor will intuitively try to maximise the effectiveness of a restricted number of measurements). Other examples are population density, rainfall, etc. The re-organisation of the data on a regular lattice is a pre-requisite for the computation of contour lines, or for various kinds of data processing. Even if the preliminary data is based on a regular lattice of sample points, a smaller lattice spacing may be required, and the problem is essentially the same. The simplest method is linear interpolation over a triangulation of the plane with the random sample points as vertices (Fig. 209). This is a two-stage process in which the z-values on the intersections of lattice edges with triangular sides are interpolated first, and then the values at lattice points. For example, if the z-value at A, B and C are known, linear interpolation gives $z(D) = (DB \cdot z(A) + AD \cdot z(B))/AB$, $z(E) = (EC \cdot z(B) + BE \cdot z(C))/BC$. Then $z(F) = (FE \cdot z(D) + DF \cdot z(E))/DE$.

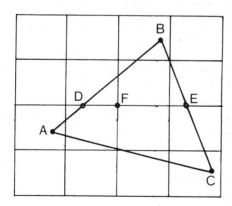

Figure 209

An alternative method of transforming randomly-sampled data to a regular lattice is to take the z-value at each lattice to be a *weighted mean* of the values at all sample points in a region surrounding the lattice point [Y1] (Fig. 210). That is,

$$z = \Sigma \, w_i z_i / \Sigma \, w_i. \tag{11.3}$$

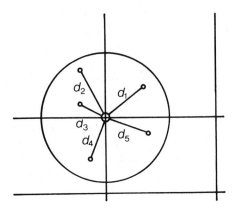

Figure 210

A reasonable choice for the weight w (a function of distance d from the lattice point) will give less weight to more distant points. The actual form of the weighting function $w(d)$ is largely arbitrary. There is little theoretical guidance for its choice. The main criterion is that it should be simple to apply. Inverse distance squared is often used. Another example, which has been successfully applied, is

$$w(d) = \begin{cases} (1 - d/R)^2/(d/R)^2, & d \leqslant R \\ 0, & d > R \end{cases} \tag{11.4}$$

The radius R of the sampling region is chosen so that about four or five sample points are included. A defect of the method is that it contains strong biases if the sample points are not fairly evenly distributed. A low-order polynomial least squares fit to the data points in a region around the lattice point is free of this defect, but is more complicated to apply.

11.2 Finite Difference Methods
The most common use of a lattice of points in a Euclidean plane or Euclidean 3-space is in the numerical solution of field equations by *finite difference* methods [S9, T1]. The method consists essentially of a simulation of the differential equations for the field by algebraic equations for a set of numbers z_{ij} providing an approximation to the values of the field at the lattice points (i, j) (in 3 dimensions, z_{ijk} representing the field at (i, j, k)).

If the differential equations are linear, the finite difference equations are also linear. An important example is that of the Laplace equation

$$\left(\frac{\partial^2}{\partial x^2} + \frac{\partial^2}{\partial y^2} \right) \phi = 0, \tag{11.5}$$

which is simulated (in the notation indicated in Fig. 211) by the set of equations (one for each lattice point of the region under consideration) of the form

$$4z_0 - z_1 - z_2 - z_3 - z_4 = 0 \qquad (11.6)$$

That is, the equations are satisfied when the z-value at each lattice point is equal to the average value at the four neighbouring points.

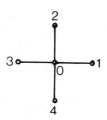

Figure 211

A difficulty that has to be overcome is the fact that the boundaries of the problem do not, in general, pass through the lattice points. A method of overcoming the difficulty (illustrated in Fig. 212) is to introduce fictitious lattice points lying beyond the boundary. For example, if the field vanishes on the boundary, a linear interpolation along the line (of unit length) joining the point 0 to the fictitious point 4 suggests that z_4 be interpreted to be $(1 - \alpha^{-1})z_0$.

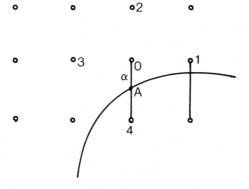

Figure 212

That is, the finite difference equation (11.6) for the point 0 should, in this case, be replaced by

$$(3 + \alpha^{-1})z_0 = z_1 + z_2 + z_3.$$

Quadratic interpolation will provide a more accurate (but computationally more complicated) treatment of the boundary. These questions are discussed in detail by Southwell [S11]. This work also contains examples of the use of regular hexagonal and equilateral triangular point systems, in place of square lattices, for the solution of Laplace's equation; also discussed is the use of changes of lattice spacing in selected regions of the plane. For example, in the case of the Laplace equation, if an extra point is introduced at the centre of a lattice square, the z-value at this extra point is to be the average of the four corner values (Fig. 213).

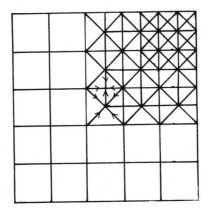

Figure 213

The *finite element* method of numerically solving field equations [M10] is based on a tessellation of the plane into regions, with considerably more freedom in the choice of shape for the regions. The method is therefore more able to cope with awkward boundary shapes.

11.3 Binary Patterns

For machine recognition of printed or written characters, the initial description of the character is a specification, for each lattice point, of whether the lattice point is inside or outside the region of the plane occupied by the character (Fig. 214). This information can be represented by a matrix z_{ij} of binary digits.

This initial description is related to the way in which the information about the form has been obtained. Information then has to be selected from it, and reorganised, to provide a description adapted to the purpose of the problem (in this case, the purpose being to identify the character). This need to transform a description in a purpose-oriented way is encountered in most problems involving form description. The information processing, in the context of character recognition (and in more general pattern-recognition problems) is known as *feature extraction*. Features are those topological and other geometrical properties of a character which, when taken together, lead to its identification. There are no rigorous criteria for the selection of features; the pro-

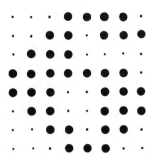

Figure 214

grammer works on the basis of intuition and experience, followed by testing the recognition process to assess its efficiency and effectiveness. We shall indicate just a few of the simpler ideas involved, chosen because they illustrate some general principles encountered in other areas of form description. For further details, the reader is referred to standard textbooks on pattern recognition, such as Duda and Hart [D6], or to Levine's review of feature extraction [L6].

A pattern like Fig. 214 is coded as a sequence of 64 binary digits a_i ($i = 1, \ldots 64$). A measure of the resemblance between two patterns A and B is provided by the quantity

$$\gamma^2 = \sum_{i=1}^{64} (a_i - b_i)^2 \qquad (11.7)$$

or, in a different notation,

$$\sigma^2 = \sum_{i=1}^{8} \sum_{j=1}^{8} (a_{ij} - b_{ij})^2 \qquad (11.8)$$

This is the lattice equivalent of the concept of variance (10.56). If a pattern A is to be compared with a *set* of patterns B, for purposes of classifying it, then the pattern B which it most resembles, according to this criterion, is the one that gives the smallest value of σ^2. Two patterns are identical if and only if σ^2 is zero. The difficulties inherent in methods based on this kind of approach come from the fact that a character may be presented in a large number of different positions and orientations. It may also occur with size variations or other more complicated distortions. Therefore a very large number of standard patterns B would have to be employed to achieve a reasonable success rate in identifying a character A.

The methods of *template matching* are similar to the above method, but they aim at identifying local features of a pattern. We illustrate the idea by a simple example. Let a_{ij} and b_{ij} be the binary matrices for two patterns, Figs. 215a, b (a_{ij} can be defined to be zero if i or j lie outside the range 1 to 8, and b_{ij} can be defined to be zero if i or j lie outside the range -1 to 1). The quantity

$$c_{ij} = \sum_k \sum_l a_{i+k,\,j+l} b_{kl} \qquad (11.9)$$

provides a comparison of each 3×3 block of the pattern A, with the 'template' B. The criterion $c_{ij} = 3$ will identify the vertical features marked in Fig. 215c, and a similar procedure with the template rotated $90°$ will pick out the horizontal features. The formula (11.9) is a lattice analogue of the concept of *convolution* (10.79), as is more apparent if it is rewritten in the form

$$c_{ij} = \sum_m \sum_n w_{i-m,\,j-n} a_{mn} \quad (w_{ij} = b_{-i,\,-j}) \qquad (11.10)$$

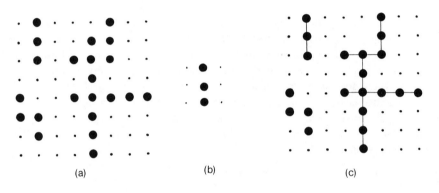

(a)   (b)   (c)

Figure 215

Elaborations of the idea make use of more complex functions of the pattern in local areas.

If the b_{ij} ($i, j = -1, 0, 1$) are all equal to 1, then (11.9) gives

$$c_{ij} = \sum_{k=i-1}^{i+1} \sum_{l=j-1}^{j+1} a_{kl}, \qquad (11.11)$$

the lattice analogue of *smoothing* by taking a moving average. It can be used for the elimination of noise from an image described on a lattice. Label the points of any chosen 3×3 block of the pattern A in the manner shown in Fig. 216. Then (11.11) is simply $c = a_1 + a_2 + \ldots a_9$. The prescription

$$a_5' = \begin{cases} 1 & \text{if } c > 5 \\ a_5 & \text{if } c = 5 \\ 0 & \text{if } c < 5 \end{cases} \qquad (11.12)$$

applied to the pattern A of Fig. 217a, will convert it to the smoothed pattern A′ of Fig. 217b. Smoothing of latticed images to eliminate noise is more useful for non-binary images, specified by a set of grey-levels.

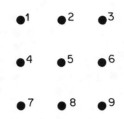

Figure 216

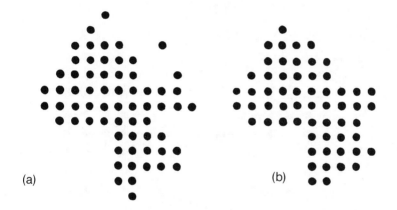

(a)	(b)

Figure 217

Pattern recognition, and more general problems of image processing, give rise to the problem of identifying the *outline* of a region of the lattice plane. One way of achieving this is by the use of 'gradient' operators. An example is

$$D = |a_1 + a_2 + a_3 - a_7 - a_8 - a_9| + |a_3 + a_6 + a_9 - a_1 - a_4 - a_7|, \quad (11.13)$$

(in the notation established in Fig. 216). The effect of applying the prescription

$$a'_5 = \begin{cases} 1 & \text{if } D \geqslant 3 \\ 0 & \text{if } D < 3 \end{cases} \quad (11.14)$$

to the pattern shown in Fig. 217a, is to convert it to the pattern in Fig. 218. These methods are more appropriate when the pattern consists of several 'grey-levels', rather than the binary patterns we are considering here. For binary

Figure 218

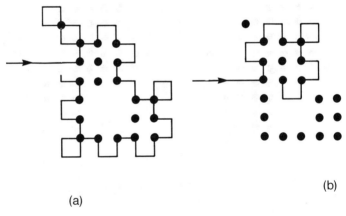

(b)

(a)

Figure 219

patterns, outlines are more effectively identified by *contour tracing* methods. A very simple example of such a method (Fig. 219a), given by Duda and Hart [D7] , is based on the following algorithm:

(i) scan the lattice until a point belonging to the character ($a_i = 1$) is encountered

(ii) If $a_i = 1$, move to the point on the left

(iii) If $a_i = 0$, move to the point on the right.

This identifies a sequence of points belonging to the boundary of the form, which can be analysed for topological and other properties, with a view to class-ification. The shortcomings of this simple algorithm are indicated in Fig. 219b. The inclusion or exclusion of a point connected only diagonally to the rest of the figure depends on the starting position of the scan; also, the contour can be led astray by getting inside an empty region within the figure. More adequate algorithms take into account the values surrounding the current position, not just the one value. These difficulties illustrate a topological peculiarity of the square lattice, associated with the concept of *connectivity*.

A point of a square lattice can be regarded either as connected to all eight neighbouring points, or alternatively, only to the four points immediately adjacent to it. Using the fourfold definition of connectivity, the 'black' region of the lattice in Fig. 220 consists of four unconnected regions, and the 'white' region consists of two unconnected regions. Thus we have a contradiction of the ordinary properties of connected regions in a continuous plane. (The Japanese board-game 'go' [H5, 12] is a fascinating exploitation of this 'paradox'). On the other hand, the 8-fold definition leads to the contradiction that the black and the white regions are *both* connected regions. To obtain consistent connectivity properties of lattice descriptions of forms, we have to use the four-fold definition for connectivity of the 'figure' and the eight-fold definition for connectivity of the 'ground', or vice-versa. This difficulty is not encountered with equilateral-triangle lattices (Fig. 140e), or with the regular point system specified by the vertices of the regular hexagonal tessellation (honeycomb). Unfortunately, computations based on these points are more complicated than those based on the regular square lattice.

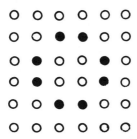

Figure 220

The general principles used in analysing digitally described patterns, that emerge from the above simple examples, are now clear. The raw data is processed to obtain it in a more easily analysable form, by smoothing to eliminate noise, identifying local features and boundaries, etc. The processed description is then analysed to identify characteristics relevant to the task of recognition, such as global topological or metric properties.

With regard to metric properties, we note that the *Euclidean metric*

$$s_E = ((i_1 - i_2)^2 + (j_1 - j_2)^2)^{\frac{1}{2}} \tag{11.15}$$

for the distance between two lattice points (i_1, j_1) and (i_2, j_2), is not very convenient, since distances between points with integer coordinates are not generally integers, and the set of lattice points within a given distance of a fixed lattice point has a rather complicated shape. Two alternative metrics, more appropriate to lattice spaces, are the *absolute value metric* (or 'city-block' metric)

$$s_A = |i_1 - i_2| + |j_1 - j_2| \qquad (11.16)$$

and the *maximum value metric*

$$s_M = \max \{|i_1 - i_2|, \ |j_1 - j_2|\} \qquad (11.17)$$

Fig. 221 shows the set of all the lattice points whose 'distance' from the origin is less than or equal to 3, for each of these three metrics.

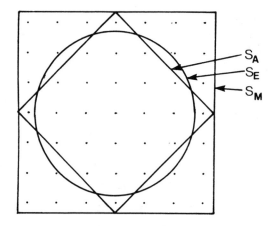

Figure 221

The description of a binary pattern on a lattice space, consisting of an array of binary digits, is the usual way of storing two-dimensional shape information in a computer. Other descriptions of the same information are possible. For example, a rectangular block of lattice points can be completely specified by giving the lattice coordinates $(i_1 \ j_1)$ and $(i_2 \ j_2)$ of its lower left and upper right vertices. This description was introduced by March and Steadman [M3], who write the coordinates as columns of a 2×2 matrix $\begin{pmatrix} i_1 & i_2 \\ j_1 & j_2 \end{pmatrix}$. For example, the pattern shown in Fig. 222 is 'the rectangle $\begin{pmatrix} 1 & 3 \\ 2 & 6 \end{pmatrix}$'. The notation for set union is then employed to describe more complicated patterns by decomposing them into component rectangles. The pattern in Fig. 223 for example, can be described as $\begin{pmatrix} 1 & 3 \\ 1 & 5 \end{pmatrix} \cup \begin{pmatrix} 0 & 4 \\ 2 & 3 \end{pmatrix} \cup \begin{pmatrix} 2 & 5 \\ 4 & 6 \end{pmatrix} \cup \begin{pmatrix} 6 & 6 \\ 6 & 6 \end{pmatrix}$. The description is not, of course, unique. Since we are using, in this example, an 8 × 8 grid, the 'usual' description requires 64 binary digits ('bits'), while the specification of the pattern in the example in the present notation requires only 48 (sixteen decimal digits requiring 3 bits each). For larger lattices, the saving can be considerable—a major criterion for the choice of a descriptive mode for computer applications is often the need to conserve memory space. A similar method based on the specification of *squares* by the coordinates of their centres, and diagonal lengths (diagonals being parallel to lattice axes), was developed by Pfalz and Rosenfeld [P1, P5] .

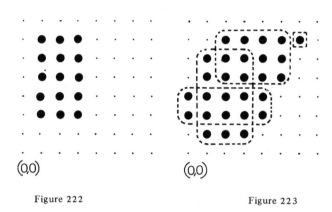

Figure 222   Figure 223

The flexibility of the method of describing shapes as unions of sets in the manner described is increased by making use also of set intersection and set complements. The possibility of employing the method for computer storage and manipulation of building plans has been explored by March and Stibbs. For buildings planned on rectangular principles (as most buildings are), it is very suitable. In this application, the rectangular regions of a rectangular lattice (with variable spacing) are labelled by integer pairs, and the lattice spacing is separately encoded, thus enabling patterns such as Fig. 224 to be encoded. See March [M5] for a more detailed discussion.

11.4 Lattice Description of Curves

A continuous curve in a Euclidean plane can be approximated by a sequence of lattice points, simply by noting those lattice points which are closest to the intersections of the curve with the lattice edges (Fig. 225). The shape of the polygonal curve so obtained can then be numerically described by the method of *chain encoding* [D7, F5, P1] , whereby the eight directions at any lattice point are given symbols (Fig. 226). The curve in Fig. 225, for example, is described by the code 222210765.

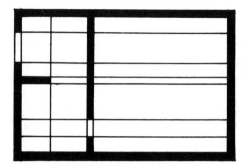

Figure 224

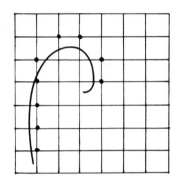 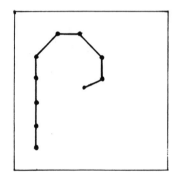

Figure 225

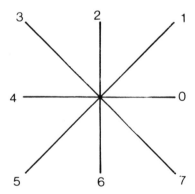

Figure 226

One of the advantages of chain codes for curves is that they are free of information about position and orientation. They are thus a discrete analogue of the intrinsic equation (2.15).

Chain encoding can also be carried out using a regular hexagonal or equilateral-triangular point system, rather than a square lattice. However, the labelling of the eight principal directions of a square lattice fully utilises the set of triples of binary digits. For hexagonal and triangular systems, the number of directions to be labelled is respectively 3 and 6. Since these numbers are not powers of two, the storage space required for the binary chain for a curve is not fully utilised. The square lattice method is therefore superior in its efficient use of computer memory space.

We conclude this chapter with a curiosity. An algorithm for generating the dragon curves (§8.1) can be formulated as a string of binary digits, in which 0 means 'turn left' and 1 means 'turn right'. Let S_N be the string for the dragon curve of order N, and let $\tilde{S}_N$ be the string obtained from it by reversing the order and changing zeros to ones, and ones to zeros. Then

$$S_1 = 1, \quad S_{N+1} = S_N 1 \tilde{S}_N \tag{11.18}$$

Chapter 12

Fourier methods

Fourier series and Fourier integrals have long been an important technique in theoretical physics, particularly in acoustics and the physics of vibration, optics, signal analysis and quantum mechanics. Our concern, of course, is with their applications in the description of geometrical form. The principal idea is the conversion of a description to an alternative, equivalent description by means of a linear operator (10.8). A conventional 'spatial' description of a complicated geometrical form is thereby transformed to a new 'frequency' description in which the form is regarded as being built up from a sequence of simpler forms by an additive process. Techniques for the analysis and manipulation of the frequency description have been used, for example, in map analysis and pattern recognition, and are of central importance in X-ray crystallography and image enhancement.

12.1 The Fourier Series

Any function of a single variable x that is *periodic* can be expressed as a linear combination of sinusoidal functions. We can, for convenience, make the period equal to 2, by a change of scale of the x-axis,

$$f(x) = f(x + 2). \tag{12.1}$$

We then have (see [C3])

$$f(x) = a_0 + \sum_{n=1}^{\infty} (a_n \cos n\pi x + b_n \sin n\pi x) \tag{12.2}$$

where the coefficients are

$$a_0 = \tfrac{1}{2}\int_{-1}^{1} f(x)\,dx$$

$$a_n = \int_{-1}^{1} f(x)\cos n\pi x \cdot dx \ (n \neq 0)$$

$$b_n = \int_{-1}^{1} f(x)\sin n\pi x \cdot dx.$$

(12.3)

The term in (12.2) corresponding to a single value of n is called a *harmonic*. Any *odd* function $f(x)$ of period 2 can be expressed as

$$f(x) = \sum_{n=1}^{\infty} b_n \sin n\pi x. \qquad (12.4)$$

Any function defined on the interval $[0, 1]$ and vanishing at $x = 0$ and $x = 1$ can be extended to form an odd periodic function, and therefore can be expressed in the form (12.4).

Applying this as a description of a standing wave on a vibrating string, fixed at its end-points $x = 0$ and $x = 1$, the time-dependence of the coefficients b_n is found by substitution in the wave equation $d^2 f/dx^2 = (1/c^2)d^2 f/dt^2$. We find that the harmonics vary sinusoidally with time, each with its own frequency. The frequency of the nth harmonic is $cn/2$.

A plane closed curve can be Fourier analysed in several ways. Its parametric equation $x = \xi(u), y = \eta(u)$ contains two functions ξ and η that are periodic in the parameter u, and so can be described in terms of their Fourier coefficients (12.3). We obtain a sequence of simplified approximations to the curve by considering only terms up to some chosen value of n. The method is of course also valid in principle, for closed space-curves. Alternatively, any closed plane curve that can be expressed as a single-valued function $r = f(\theta)$ can be expressed as a Fourier series (replacing πx by θ in (12.2)). The first few harmonics, alone and with an a_0 term added, are illustrated in Fig. 227a, and a curve with a particularly simple Fourier series $(1 + \tfrac{3}{4}\cos\theta + \tfrac{1}{2}\cdot\cos 2\theta)$ is illustrated in Fig. 227b. A third possibility is the Fourier analysis of the curvature $\kappa(s)$ of a closed curve, which is of course a periodic function of arc length s. However, this method is useless for curves with 'corners', where the curvature is infinite. The angle

$$\psi = \int_{0}^{s} \kappa(s)\,ds \qquad (12.5)$$

which the tangent makes with the tangent at the starting point, is not periodic $(\psi(s + L) = \psi(s) + 2\pi$, where L is the total length of the curve). However, $\psi(s) - 2\pi s/L$ is periodic, and the function

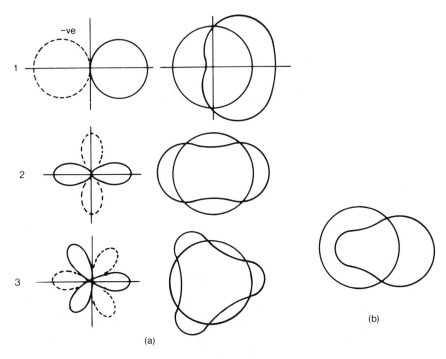

Figure 227

$$\hat{\theta}(s) = \psi(s) - \frac{2\pi s}{L} - \mu, \quad \mu = \frac{1}{L} \int_0^L \left(\theta(s) - \frac{2\pi s}{L} \right) ds \qquad (12.6)$$

is periodic *and* independent of the choice of the starting point $s = 0$. The description based on the Fourier analysis of $\hat{\theta}(s)$ was introduced by Brill [B9]. His results, obtained by truncating the Fourier series, are shown in Fig. 228.

A convenient form for the Fourier series (12.2) is obtained with the aid of the complex function

$$e^{i\theta} = \cos\theta + i\sin\theta \qquad (12.7)$$

Defining $c_0 = a_0$, $c_n = \frac{1}{2}(a_n - ib_n)$, $c_{-n} = \frac{1}{2}(a_n + ib_n)$, $(n \geqslant 1)$, we have

$$f(x) = \sum_{n=-\infty}^{\infty} c_n e^{in\pi x} \qquad (12.8)$$

where the complex coefficients are given by

$$c_n = \frac{1}{2} \int_{-1}^{1} f(x) e^{-in\pi x} dx \qquad (12.9)$$

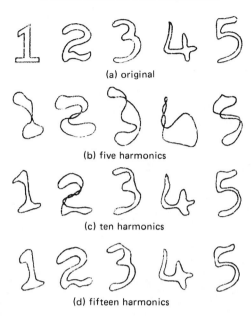

(a) original

(b) five harmonics

(c) ten harmonics

(d) fifteen harmonics

Figure 228

12.2 Fourier Series in Two Dimensions

If a function of two variables $f(x, y)$ is periodic in both of them, the Fourier expansion can be applied to both. By changing the scale of both variables so that the periods are both equal to 2, we get the double Fourier series

$$f(x, y) = \sum_{n=-\infty}^{\infty} \sum_{m=-\infty}^{\infty} c_{nm}\, e^{i\pi(nx + my)} \qquad (12.10)$$

For example, any function on the interior of a unit square with vertices (00) (01) (11) and (10) in the Cartesian coordinate system (x, y), which vanishes on the perimeter of the square, can be written in the form

$$f(x, y) = \sum_{n=-\infty}^{\infty} \sum_{m=-\infty}^{\infty} b_{nm}\, (\sin n\pi x \sin m\pi y), \qquad (12.11)$$

If (12.11) is applied as a description of the form of a standing wave $z = f(x, y, t)$ on a vibrating square membrane, the time dependence of the coefficients b_{nm} is obtained by substitution in the wave equation

$$\left(\frac{\partial^2}{\partial x^2} + \frac{\partial^2}{\partial y^2} \right) f = \frac{1}{c^2} \frac{\partial^2}{\partial t^2} f. \qquad (12.12)$$

We find that the harmonic associated with b_{nm} varies sinusoidally with time, with frequency

$$v = \frac{c}{2} \sqrt{m^2 + n^2}.$$ (12.13)

Some of the simpler harmonics for a vibrating square plate are illustrated in Fig. 229 (the dotted contours refer to negative z). Fig. 230 shows the appearance of a membrane vibrating in a mode in which just the two harmonics with m, n equal to 1, 2 and 2, 1 are excited, with equal amplitude.

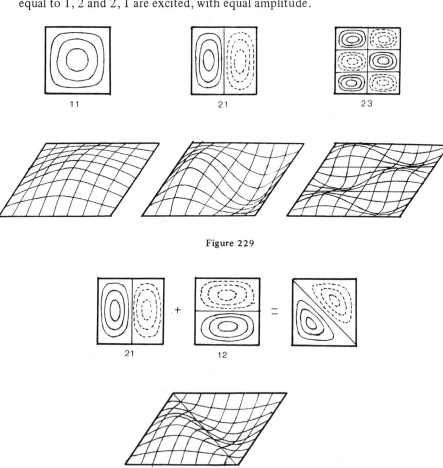

Figure 229

Figure 230

The reason why the description (12.11) turned out to be particularly appropriate for the form of a vibrating square membrane is that the functions

$$\phi_{nm} = \sin n\pi x \sin m\pi y$$ (12.14)

are the eigenfunctions of the Laplacian operator

$$\left(\frac{\partial^2}{\partial x^2} + \frac{\partial^2}{\partial y^2}\right) \tag{12.15}$$

appearing in the wave equation, determined by the particular boundary conditions. For a differently-shaped membrane, a different set of eigenfunctions of (12.15) would have been appropriate. For example,

$$\phi_{nm} = J_m(k_n r) \begin{cases} \sin m\pi x \\ \cos m\pi x \end{cases} \tag{12.16}$$

are the eigenfunctions of the Laplacian that vanish on the boundary of the unit circle, where J_m is a Bessel function and k_n is its nth zero $(J_m(k_n) = 0)$. The harmonic $n = 3$, $m = 4$ for a circular membrane, for example, has the appearance of Fig. 231. Any function defined on the interior of the unit circle, that vanishes on its circumference, can be expressed as a linear combination of the functions (12.16). Unfortunately, the eigenvalues of the Laplacian for more complicated shapes cannot usually be obtained in analytic form, and the method loses its usefulness.

Any function $f(\theta, \varphi)$ defined on the surface of a sphere can be expanded as a linear combination of the spherical harmonics [B10, M13] $Y_{lm}(\theta, \varphi)$ $(l = 0, 1, 2 \ldots ; m = -l, -l + 1, \ldots l)$:

$$f(\theta, \varphi) = \sum_{l=0}^{\infty} \sum_{m=-l}^{\infty} c^{lm} Y_{lm}(\theta, \varphi), \tag{12.17}$$

$$c^{lm} = \oint f Y_{lm}^* \sin\theta \, d\theta \, d\varphi \tag{12.18}$$

where the integration in (12.18) is over the whole surface of the sphere. The spherical harmonics are defined to be the eigenvalues of

$$\Lambda = \frac{1}{\sin\theta}\left(\frac{\partial}{\partial\theta}\left(\sin\theta \frac{\partial}{\partial\theta}\right) + \frac{1}{\sin\theta}\frac{\partial^2}{\partial\varphi^2}\right) \tag{12.19}$$

and of $\partial/\partial\varphi$. Specifically,

$$\left.\begin{array}{l} \Lambda Y_{lm} = -l(l-1) Y_{lm} \\[2mm] \dfrac{\partial}{\partial\varphi} Y_{lm} = im Y_{lm} \end{array}\right\} \tag{12.20}$$

The operator Λ is the angular part of the 3-dimensional Laplacian:

$$\Delta^2 = \frac{\partial^2}{\partial x^2} + \frac{\partial^2}{\partial y^2} + \frac{\partial^2}{\partial z^2} = \frac{1}{r^2}\left(\frac{\partial}{\partial r} r^2 \frac{\partial}{\partial r} + \Lambda\right). \tag{12.21}$$

Figure 231

The first few spherical harmonics are given by

$$Y_{00} = \frac{1}{\sqrt{4\pi}}$$

$$Y_{10} = \sqrt{\frac{3}{4\pi}} \cos\theta \qquad\qquad Y_{11} = \sqrt{\frac{3}{8\pi}} \sin\theta \, e^{i\varphi} \qquad\qquad (12.22)$$

$$Y_{20} = \sqrt{\frac{5}{4\pi}} \frac{1}{2}(3\cos^2\theta - 1) \quad Y_{21} = \sqrt{\frac{15}{8\pi}} \sin\theta \cos\theta \, e^{i\varphi} \quad Y_{22} = \frac{1}{4}\sqrt{\frac{15}{2\pi}} \sin^2\theta \, e^{2\varphi},$$

those with negative m being given by the prescription

$$Y_{l-m} = (-)^m \, Y_{lm}^*. \qquad\qquad (12.23)$$

A few of the spherical harmonics are illustrated in Fig. 232 (the real forms $\frac{1}{2i}(Y_{11} + Y_{1-1})$, $\frac{1}{2i}(Y_{21} - Y_{2-1})$ and $\frac{1}{2i}(Y_{22} - Y_{2-2})$ are related to the given forms $\frac{1}{2}(Y_{11} - Y_{1-1})$, $\frac{1}{2}(Y_{21} + Y_{2-1})$ and $\frac{1}{2}(Y_{22} + Y_{2-1})$ by a rotation of $90°$ about the z-axis). The illustrations are of the surfaces $r = Y(\theta, \varphi)$.

The 'multipole' expansion (12.7) occurs in the analysis of the vibrational modes of a sphere [S13], in the description of radiation from a point source [L3], in the description of the shape of a deformed sphere such as the earth [S13], in the analysis of the gravitational and magnetic fields of the earth [S13] and in the shapes of the electron distribution of a hydrogen atom [E3, S2]. In illumination engineering, the scalar illumination and the illumination vector [G3, L10] are the $l = 0$ and $l = 1$ parts of the illumination distribution function.

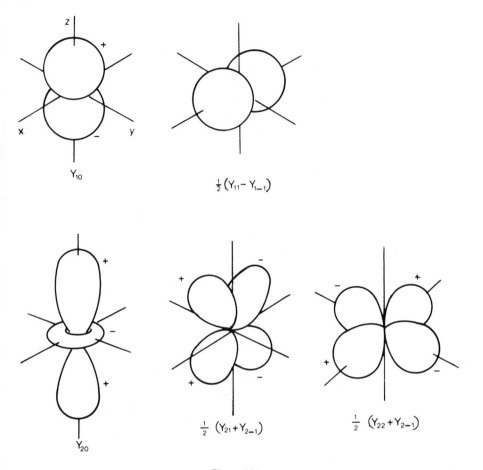

Figure 232

12.3 Walsh Functions

Orthonormal sets of functions other than the eigenfunctions of the Laplacian operator can be used for a Fourier analysis of a function. An important example from information theory is the set of Walsh functions [H4]. These are defined on the interval $(\frac{1}{2}, \frac{1}{2})$ (and, outside this region, by the requirement that they be periodic of period 1). To define them, we have to introduce an operation $\oplus$ for combining a pair of integers m and n; $m \oplus n$ is defined by writing m and n in binary notation and then adding their digits using modulo 2 arithmetic. For example, $22 \oplus 13 = 10110 \oplus 01101 = 11011 = 27$. The Walsh functions $W_l(x)$

($l = 0, 1, 2 \ldots$) are then defined by

$$W_1(x) = \text{sign}(\sin \pi x)$$
$$W_0 = 1 \qquad W_{2^n}(x) = \text{sign}(\cos 2^n \pi x)\,(n > 0) \left.\right\} \qquad (12.24)$$
$$W_{m \oplus n}(x) = W_m(x)\,W_n(x)$$

For example, $W_7 = W_{4 \oplus 2 \oplus 1} = W_4 W_2 W_1 = \text{sign}(\cos 4\pi x \cdot \cos 2\pi x \cdot \sin \pi x)$. Any Walsh function, at any value of x, is always equal to $+1$ or to -1. The first few are shown in Fig. 233.

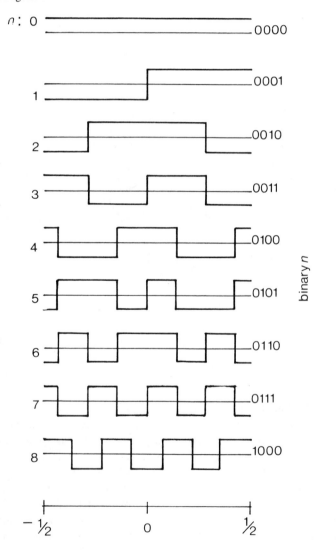

Figure 233

A function $f(x, y)$ defined on the unit square (vertices $(\pm\frac{1}{2}, \pm\frac{1}{2})$) in terms of the coefficients of its expansion in two-dimensional Walsh functions:

$$f(x, y) = \sum_{i=0}^{\infty} \sum_{j=0}^{\infty} c_{ij} W_{ij}(x, y), \qquad (12.25)$$

$$c_{ij} = \int_{-\frac{1}{2}}^{\frac{1}{2}} \int_{-\frac{1}{2}}^{\frac{1}{2}} f(x, y) W_{ij}(x, y) \, dx \, dy, \qquad (12.26)$$

where

$$W_{ij}(x, y) = W_i(x) W_j(y). \qquad (12.27)$$

The Walsh function $W_{2,5}$ for example, is shown in Fig. 234. Shaded areas and blank areas refer, respectively, to the values $+1$ and -1.

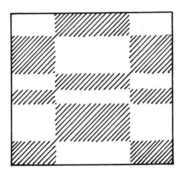

Figure 234

Meltzer, Searle and Brown [M8] have applied this analysis to binary functions $f(x, y)$ describing leaf forms (i.e., $f(x, y)$ take the value 1 when (x, y) is within the leaf boundary, otherwise it is zero). The idea is that intricate outlines will be associated with large high-order coefficients in (2.25). The method does not appear to be particularly appropriate. Any function defined on the unit square can be expanded in terms of Walsh functions W_{ij}; binary functions have no special significance in this respect. Several hundred terms were found to be required to obtain reasonable approximations to the original leaf shapes. D'Arcy Thompson's observation [T3] that simple sinusoidal functions (Fig. 235) look like leaf forms seems to indicate that a more conventional Fourier analysis of leaf boundaries, like Brill's method [B9], might be a more fruitful approach to classification.

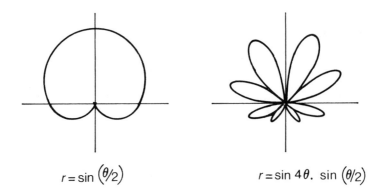

$r = \sin\left(\theta/2\right)$ $r = \sin 4\theta.\ \sin\left(\theta/2\right)$

Figure 235

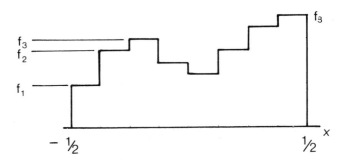

Figure 236

Walsh function Fourier analysis is particularly suited for functions $f(x)$ which are stepwise discontinuous functions on 2^n equal intervals (e.g. Fig. 236). The Fourier coefficient

$$c_3 = \int_{-\frac{1}{2}}^{\frac{1}{2}} f(x)\, W_3(x)\, dx \qquad (12.28)$$

for such a function, for example, is simply $f_1 + f_2 - f_3 - f_4 + f_5 + f_6 - f_7 - f_8$. The coefficients c_{ij} for a numerically described picture function (e.g. Fig. 208a) defined on a $2^n \times 2^n$ lattice would be similarly obtained simply by addition and subtraction of 'grey-values', and all the c_{ij} with i or j greater than 2^n would vanish.

12.4 The Fourier Integral
Rewrite the Fourier series (12.8) and the prescription (12.9) for its coefficients in terms of $g(x) = f(x/L)$, $\omega = n\pi/L$ and $G(\omega) = L c_n/\pi$. We get

$$g(x) = \sum_{n=-\infty}^{\infty} \frac{\pi}{L} G(\omega) e^{i\omega x}, \tag{12.29}$$

$$G(\omega) = \frac{1}{2\pi} \int_{-L}^{L} g(x) e^{-i\omega x} dx \tag{12.30}$$

Taking the limit $L \to \infty$, the summation over n in (12.29) becomes an integration over ω. The result is the *Fourier integral* [C1] expression for a function $g(x)$, which generalises the idea of the Fourier series to functions that are not periodic:

$$g(x) = \int_{-\infty}^{\infty} G(\omega) e^{i\omega x} d\omega \tag{12.31}$$

$$G(\omega) = \frac{1}{2\pi} \int_{-\infty}^{\infty} g(x) e^{-i\omega x} dx \tag{12.32}$$

The function $G(\omega)$ defined by (12.32) is the Fourier *transform* of $g(x)$.

To see how this can lead to the discrete series, in the case where $g(x)$ *is* periodic, suppose that $g(x) = g(x + 2)$. Then

$$G(\omega) = \frac{1}{2\pi} \int_{-\infty}^{\infty} g(x + 2) e^{-i\omega x} dx.$$

Change the variable to $y = x + 2$, and we find

$$G(\omega) = e^{2i\omega} G(\omega). \tag{12.33}$$

Therefore $G(\omega)$ is zero except when ω has the form $n\pi$, where n is an integer. Such a function can be expressed in terms of the delta-function defined in §10.8:

$$G(\omega) = \sum_{n=-\infty}^{\infty} a_n \delta(\omega - n\pi). \tag{12.34}$$

Substituting this expression in the Fourier integral (12.31), we find that the integration can be done because of the properties of the delta-function. Namely, because of (10.72),

$$\int_{-\infty}^{\infty} \delta(\omega - n\pi) e^{i\omega x} d\omega = e^{in\pi x} \tag{12.35}$$

Thus, the Fourier integral for a function $g(x)$ of period 2 reduces to the usual discrete Fourier *series*,

$$g(x) = \sum_{n=-a}^{\infty} a_n e^{in\pi x}. \tag{12.36}$$

For a function $g(x, y)$ of two variables, the double Fourier series (12.10) generalises to the double Fourier integral

$$g(x) = \int_{-\infty}^{\infty} \int_{-\infty}^{\infty} G(\boldsymbol{\omega}) e^{i\boldsymbol{\omega} \cdot x} d\omega_1 \, d\omega_2 \qquad (12.37)$$

$$G(\boldsymbol{\omega}) = \frac{1}{(2\pi)^2} \int_{-\infty}^{\infty} \int_{-\infty}^{\infty} g(x) e^{-i\boldsymbol{\omega} \cdot x} dx dy \qquad (12.38)$$

where we have written x for the position vector in the (x, y)-plane and $\boldsymbol{\omega}$ for the position vector in the (ω_1, ω_2) plane.

For a *real* function $g(x)$, we must have $G^*(\boldsymbol{\omega}) = G(-\boldsymbol{\omega})$. Now,

$$z = A(\boldsymbol{\omega}) e^{i\boldsymbol{\omega} \cdot x} + A^*(\boldsymbol{\omega}) e^{-i\boldsymbol{\omega} \cdot x} \qquad (12.39)$$

is a real sinusoidal surface with wavelength $2\pi/\omega (\omega = \sqrt{\omega_1^2 + \omega_2^2})$, with amplitude and phase given by the modulus and argument of $A(\boldsymbol{\omega})$ (Fig. 237). Therefore the Fourier integral can be regarded as the description of a surface $z = g(x)$ as a superposition of an infinite number of sinusoidal surfaces. We expect small-scale variations of the surface to be associated with large values of ω, and the more general 'smoothed out' aspects of the surface to be associated with small values of ω. The use of the Fourier transformation in cartography [R4] is based on this effect. If $g(x)$ is a 'picture function' describing an image in terms of grey-values, then the image quality can be changed by suppressing some values of ω and enhancing others. This is the basic concept behind the idea of *filtering* in computer processing of images [D7, R5].

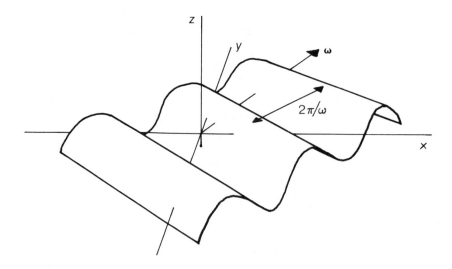

Figure 237

Figure 238

Suppose now that $g(x)$ is symmetrical under one of the seventeen 'wallpaper' groups (§8.5). That is, suppose it forms a regular pattern over the (x, y)-plane as in Fig. 238, which represents by contours the electron density in a plane through a crystal of diopside, taken from Battey [B1]. Every transformation of the group can be written in the form

$$x' = Ax + a \qquad (12.40)$$

where A is a 2×2 orthogonal matrix. The pattern satisfies

$$g(x) = g(Ax + a) \qquad (12.41)$$

for every transformation that belongs to the group. It follows, from substituting this in the integral (12.38) and changing the Cartesian axes, that

$$G(\omega) = e^{i(A\omega) \cdot a} G(A\omega) \qquad (12.42)$$

For example, for the translations $x' = x + a$ that leave the pattern unchanged, $G(\omega)$ must be zero except when

$$\omega \cdot a = 2N\pi \qquad (12.43)$$

where N is an integer. Now, we already saw that the sinusoidal component of $g(x)$ associated with the vector ω had wavelength $\lambda = 2\pi/\omega$, so we can write $\omega = (2\pi/\lambda)n$, n being the unit vector in the direction of the sinusoid. Thus, the sinusoidal components of $g(x)$ must satisfy

$$n \cdot a = N\lambda \qquad (12.44)$$

for some integer N, for *every* lattice translation a. This simply means that the sinusoidal components must 'fit into' the lattice as in Fig. 239. Since *every* lattice translation a is of the form $N_1 a_1 + N_2 a_2$ where a_1 and a_2 are the generators of the translation subgroup (i.e., a_1 and a_2 are vectors along the sides of a

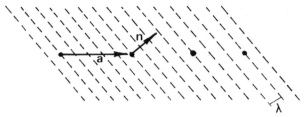

Figure 239

primitive parallelogram of the lattice) and N_1 and N_2 are integers, (12.43) can only be satisfied for *all* lattice translations if

$$\left.\begin{array}{l} \boldsymbol{\omega} \cdot \boldsymbol{a}_1 = 2\pi n \\ \boldsymbol{\omega} \cdot \boldsymbol{a}_2 = 2\pi m \end{array}\right\} \tag{12.45}$$

with integer n and m. If we choose oblique axes so that $\boldsymbol{a}_1$ and $\boldsymbol{a}_2$ are $(2, 0)$ and $(0, 2)$, then

$$\left.\begin{array}{l} \omega_1 = n\pi \\ \omega_2 = m\pi \end{array}\right\} \tag{12.46}$$

for $G(\boldsymbol{\omega})$ to be non-zero. Hence $G(\boldsymbol{\omega})$ can be expressed in terms of the delta-functions.

$$G(\boldsymbol{\omega}) = \sum_{n=-\infty}^{\infty} \sum_{m=-\infty}^{\infty} c_{nm}\, \delta\,(\omega_1 - n\pi)\, \delta\,(\omega_2 - n\pi).$$

Substituting this into the Fourier integral (12.37) and carrying out the integration, we find that we have recovered the *series* (12.10) for a doubly-periodic function. The analysis for the three-dimensional case (for a function $g(x, y, z)$ symmetrical under a space-group, such as electron density in a crystal) is exactly analogous. Application of (12.41) for the point symmetries, glide reflections and rotatory translations lead to symmetry conditions to be satisfied by the coefficients in the Fourier integral. The triple Fourier series is fundamental to the techniques of X-ray crystallography [W11].

12.5 Bandlimiting
The Fourier transform of a function (10.78) that is zero outside the interval $\left(-\dfrac{W}{2}, \dfrac{W}{2}\right)$ and equal to 1 within the interval is

$$G(\omega) = \frac{1}{2\pi} \int_{-\frac{W}{2}}^{\frac{W}{2}} e^{i\omega x} dx = \frac{W}{2} \operatorname{sinc}\left(\frac{\omega W}{2}\right) \tag{12.47}$$

where the function sinc is defined by

$$\text{sinc}(\theta) = \frac{\sin\theta}{\theta} \qquad (12.48)$$

The 'box car' function and its Fourier transform are shown in Fig. 240.

Note that the central peak of $G(\omega)$ becomes *lower and wider as W is decreased, higher and narrower as W is increased* – an illustration of the observation that small scale features of a function are associated with the higher values of ω in its transform.

A function $g(x)$ for which the transform $G(\omega)$ is zero above a certain 'frequency' (say, $G(\omega) = 0$ for $|\omega| > K$) is said to be *bandlimited*. The *Shannon sampling theorem* [D6] states that all the information contained in a bandlimited function $g(x)$ is preserved if it is sampled at a set of discrete points $x = n\pi/K$. The proof is as follows. For a bandlimited function $g(x)$,

$$g(x) = \int_{-K}^{K} G(\omega)e^{-i\omega x} \qquad (12.49)$$

The Fourier *series* for $G(\omega)$ will be valid in the interval $(-K, K)$. It is

$$G(\omega) = \sum_{n=-\infty}^{\infty} c_n e^{in\pi\omega/K}, \qquad (12.50)$$

where

$$c_n = \frac{1}{2K}\int_{-K}^{K} G(\omega)e^{-in\pi\omega/K}d\omega = \frac{1}{2K} g\left(\frac{n\pi}{K}\right). \qquad (12.51)$$

Substituting in (12.50) and carrying out the integration, we get

$$g(x) = \sum_{n=-\infty}^{\infty} g\left(\frac{n\pi}{K}\right)\text{sinc}(n\pi - Kx). \qquad (12.52)$$

This shows that a bandlimited function is completely determined by its value at a discrete set of points $x = n\pi/K$. The continuous description is recovered completely from the sampled values by a superposition of weighted sinc-functions centred at the sample points. In the two-dimensional generalisation, we conclude that, if the Fourier transform of a function $g(x, y)$ (eg. a 'picture function') vanishes outside a rectangle of sides $2K_1$ and $2K_2$, centred at the origin of the (ω_1, ω_2) plane, then the continuous description can be recovered from the description given by sampling on a lattice with spacing $2\pi/K_1$ along the x-direction and $2\pi/K_2$ along the y-direction; no information is lost by the sampling procedure.

A bandlimited function $g(x)$ has the following property: if $|g(x)| < M$ for all x, and the band-width is K, then

$$|dg/dx| < 2\pi KM \qquad (12.53)$$

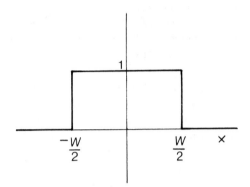

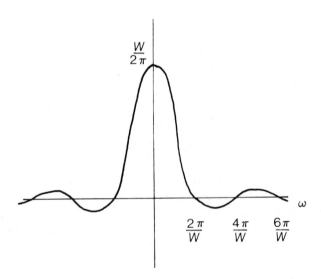

Figure 240

(Bernstein's inequality [D6]). Thus, one can establish criteria for deciding on the optimum spacing of sample points in order that a discrete description of a surface $z = g(x,y)$ shall contain essential information, based on a knowledge of the way the height and the gradient are varying.

When a general function $g(x)$ is smoothed by discarding higher 'frequencies' in its Fourier transform (say for $|\omega| > K$), then the smoothed function is the function whose Fourier transform is

$$H(\omega)\, G(\omega),$$

H being the function that is equal to 1 in the interval $[-K, K]$ and zero outside this interval. If we let H be a more general function (satisfying $H^*(\omega) = H(-\omega)$), we have the concept of *filtering*. Various frequencies can be repressed or enhanced according to the choice of H. In particular, random error in $g(x)$ (noise) with a characteristic frequency spectrum can be suppressed. The effect of filtering on a function $g(x)$ is deducible from the theorem that *the product of two Fourier transforms is the Fourier transform of the convolution of the two associated functions*. Thus, if $h(x)$ is the function whose transform is $H(\omega)$, the effect of the filtering is to convert $g(x)$ to

$$(g * h) = \int_{-\infty}^{\infty} h(x - u)\, g(u)\, du \qquad (12.54)$$

Thus, the filtering process corresponds to taking a moving weighted average with $h(-x)$ as the weighting function. In the case of smoothing by a simple band-limiting, $g(x)$ is smoothed by convolution with a sinc function $2K$ sinc Kx. The effect of this on the $g(x)$ of Fig. 240 is shown in Fig. 241. Thus, bandlimiting of a picture function will soften the edges of discontinuities, and will also give them a 'banded' appearance (the Gibbs phenomenon). The analogue of bandlimiting, in the (x,y)-plane, is the limiting of a function $g(x,y)$ to a region about the origin, setting it to zero outside the region. In practical applications, of course, this is the kind of function we are dealing with (for example, an image or a geographical region); the region to which the information is restricted is almost always a rectangle. A topographic map function $g(x)$ is characterised by a high peak at the origin of the (ω_1, ω_2)-plane in its Fourier transform $G(\boldsymbol{\omega})$, cor-

Figure 241

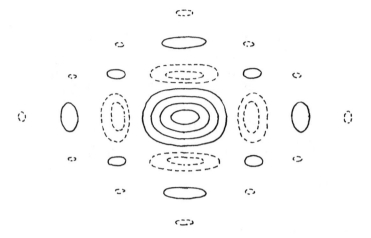

Figure 242

responding to the broad general features of the topography. Discarding of information outside a rectangular region of the (x, y)-plane changes $G(\omega)$ by convoluting it with the function sinc $(\omega_1 W_1/2)$ sinc $(\omega_2 W_2/2)$, so the central peak in $G(\omega)$ becomes a pattern whose contours are indicated in Fig. 242. The point here is that the theoretic approach to Fourier analysis deals with functions $g(x)$ and $G(\omega)$ which are defined over the *whole* of their respective spaces, whereas in practical applications one has to deal with finite regions, and this imposes limitations on the use of Fourier transforms as a descriptive technique.

12.6 The Discretised Fourier Transform
To evaluate an integral numerically, it has to be approximated by a summation over a finite number of terms. However, the straightforward discretisation of (12.31) and (12.32) gives rise in general to (matrix) formulae which are not mutually inverse. The *fast Fourier transform* is a special kind of numerical approximation to (12.31) and (12.32) which is free of this defect, and which incorporates some clever tricks which drastically reduce the amount of computation required. The method has therefore become an indispensible tool in any enterprise involving massive amounts of computation of Fourier integrals, such as picture processing.

The details of the fast Fourier transform are outside the scope of our subject matter. The interested reader will find them clearly presented by Brigham [B8].

Bibliography and References

A1. Ahlberg, J. H., Nilson, E. N. and Walsh, J. L., *The theory of splines and their applications*, Academic Press, New York (1967).

A2. Albert, A. A. and Sandler, R., *An introduction to finite projective planes*, Holt, Reinhart and Winston (1968).

B1. Battey, M. H., *Mineralogy for students*, Oliver and Boyd (1972).

B2. Baybars, I. and Eastman, C. M., Enumerating architectural arrangements by generating their underlying graphs, *Environment and Planning b*, 7 (1980) 289-310.

B3. Blum, H., A model for extracting new descriptors of shape (ref. W6) (1967) 362-380.

B4. Blum, H., A geometry for biology, *Math. analysis of fundamental biological phenomena*, 231 (no. 1) (1974) 19-30.

B5. Bourgoin, J., *Arabic geometrical pattern and design*, Firmin-Didot, Paris (1879), Dover (1973).

B6. Bragg, W. L., *Atomic structure of minerals*, Oxford University Press (1937).

B7. Bragg, W. L. and Claringbull, G. F., *The crystal structure of minerals*, G. Bell, London (1965).

B8. Brigham, E. O., *The fast Fourier transform*, Prentice-Hall (1974).

B9. Brill, E. L., *Character recognition via Fourier descriptors*, Wescon paper 25/3 (1968).

B10. Brink, D. M. and Satchler, G. R., *Angular momentum*, Oxford University Press (1962).

B11. Brückner, M., *Vielecke und Vielfläche, Theorie und Geschichte*, Teubner, Leipzig (1900).

B12. Burgess, C. E. and Cannon, J. W., Embedding of surfaces in E_3, *Rocky Mountain J. Math.*, 1 (no. 2) (1971) 259-344.

B13. Burks, A. W. (ed.), *Essays on cellular automata*, University of Illinois (1970).

C1. Champeney, D. C., *Fourier transforms and their applications*, Academic Press (1973).

C2. Cheng, G. S., Ledley, R. S., Pollock, D. K. and Rosenfeld, A. (eds.), *Pictorial pattern recognition*, Thompson, Washington (1968).

C3. Churchill, R. V., *Fourier series and boundary value problems*, McGraw-Hill (2nd edn, 1969).

C4. Clark, K., *Civilization*, BBC (1967) 187.

C5. Conte, S. D. and De Boor, C., *Elementary numerical analysis*, McGraw-Hill Kogakusha (2nd edn, 1972).

C6. Coolidge, J. L., *Algebraic plane curves*, Oxford University Press (1931).

C7. Coolidge, J. L., *A history of the conic sections and quadric surfaces*, Oxford University Press (1945).

C8. Coolidge, J. L., *A history of geometrical methods*, Dover (1963).

C9. Coons, S. A., Surfaces for the computer aided design of space figures, M.I.T. ESL 9442-M-139, Jan. (1964).

C10. Coons, S. A., Surfaces for the computer aided design of space forms, M.I.T. MAC-TR-41, Jan. (1967).

C11. Coxeter, H. S. M., Regular honeycombs in hyperbolic space, *Proc. Int. Congress of Mathematicians III* (1954) 155-169.

C12. Coxeter, H. S. M., *The real projective plane*, Cambridge University Press (2nd edn, 1955).

C13. Coxeter, H. S. M., *Non-Euclidean geometry*, University of Toronto Press (4th edn, 1961).

C14. Coxeter, H. S. M., *Geometry*, Wiley (2nd edn, 1969).

C15. Coxeter, H. S. M., Angels and devils (ref. K4) (1981) 197-209.

C16. Critchlow, K., *Order in space*, Thames and Hudson (1969).

C17. Critchlow, K., *Islamic patterns*, Thames and Hudson (1976).

D1. Davis, H. T., *Introduction to nonlinear differential and integral equations*, United States Atomic Energy Commission (1960), Dover (1962).

D2. Davis, J. C., *Statistics and data analysis in geology*, Wiley (1973).

D3. Davis, J. C. and McCullagh, M. J. (eds.), *Display and analysis of spatial data*, Wiley International (1975).

D4. Delfiner, P. and Delhomme, J. P., Optimum interpolation by Kriging (ref. D3) (1975) 96-114.

D5. Do Carmo, M. P., *Differential geometry of curves and surfaces*, Prentice-Hall (1976).

D6. Duda, R. O. and Hart, P. E., *Pattern classification and scene analysis*, Wiley (1973).

D7. Duncan, J. P. and Vickers, G. W., Simplified method for interactive adjustment of surfaces, *Computer Aided Design*, 12 (no. 6) (1980) 305-308.

D8. Düppe, R. D. and Gottschalk, H. J., Automatische Interpolation von Isolinien bei willkurlich verteilten stützpunkten, *Allgemeine Vermessungs-nachrichten*, 77 (1970) 423-426.

E1. Edge, W. L., *The theory of ruled surfaces*, Cambridge University Press (1931).

E2. Eells, J., *Singularities of smooth maps*, Gordon and Breach (1967).
E3. Eisberg, R. and Resnick, R., *Quantum physics of atoms, molecules, solids, nuclei and particles*, Wiley (1974).
E4. El-Said, I. and Parman, A., *Geometric concepts in Islamic art*, World of Islam Festival Publishing Company (1976).
E5. Engel, H., *Structure systems*, Iliffe Books, London (1968).
F1. Fejes Tóth, L., *Regular figures*, Pergamon (1964).
F2. Flügge, W., *Stresses in shells*, Springer (2nd edn, 1973).
F3. Forrest, A. R., On Coons and other methods for the representation of curved surfaces, *Computer Graphics and Image Processing*, 1 (1972) 341-359.
F4. Forsyth, A. R., *Calculus of variations*, Dover (1960).
F5. Freeman, H., On the encoding of arbitrary geometric configurations, *Inst. Radio Engineers Trans. Electronic Computers, ECIO* (1961) 260-268.
G1. Gardner, M., Mathematical games, *Scientific American*, Mar. (1967) 125.
G2. Gardner, M., Mathematical games, *Scientific American*, Jan. (1977) 110.
G3. Gershun, A., The light field, *J. Maths. Phys.*, 18 (1939) 51-151.
G4. Gips, J., *Shape grammars and their uses*, Birkhäuser (1975).
G5. Golubitsky, M. and Guillemin, V., *Stable mappings and their singularities*, Springer (1973).
G6. Grant, F., A problem in the analysis of geophysical data, *Geophysics*, 22 (1957) 309-344.
G7. Groner, G. F., Real-time recognition of handprinted text, *RAND Memorandum* RM-5016-ARPA (1966).
G8. Grünbaum, B. and Shephard, G. C., Some problems on plane tilings (ref. K4) (1981) 167-196.
H1. Hahn, H., Geometry and intuition, *Scientific American*, Apr. (1954) 84.
H2. Harmon, L. D., The recognition of faces, *Scientific American*, Nov. (1973) 71-82.
H3. Harmon, L. D. and Julesz, B., Masking in visual recognition, effect of two-dimensional filtered noise, *Science*, 180 (no. 4091) (1973).
H4. Harmuth, H. F., *Transmission of information by orthogonal functions*, Springer (1969).
H5. Haruyama, I. and Nagahara, Y., *Basic techniques of go*, Ishi Press, Tokyo (1980).
H6. Hausmann, J. C. (ed.), *Knot theory*, Springer (1978).
H7. Heaton, P., *Sailing*, Penguin (3rd edn, 1966).
H8. Hilbert, D. and Cohn-Vossen, S., *Geometry and the imagination*, Chelsea Publishing Co., New York (1952).
H9. Hilditch, J., An application of graph theory in pattern recognition, *Machine Intelligence Vol. 3* (D. Mitchie, ed.), Edinburgh University Press (1968).
H10. Hinks, A. R., *Map projections*, Cambridge University Press (1921).
H11. Homegraaf, A., Zonoids and zonoidal complexes, *World Congress on*

Space Enclosures I, Building Research Centre, Concordia Univ., Montreal (1976) 73-82.

H12. Horton, R. E., Erosional development of streams and their drainage basins: hydrophysical approach to quantitative morphology, *Bull. Geog. Soc. Amer.*, **56** (1945) 275-370.

H13. Hsu, M. L., Filtering process in surface generalisation and isopleth mapping (ref. D3) (1975) 115-129.

H14. Hsu, M. L. and Robinson, A. H., *The fidelity of isopleth maps, an experimental study*, University of Minnesota Press (1970).

H15. Hu, M. K., Visual pattern recognition by moment invariants, *IRE Trans. Information Theory, IT-8*, Feb. (1962) 179-187.

H16. Huijbregts, C. J., Regionalised variables and quantitative analysis of spatial data (ref. D3) (1975) 38-53.

H17. Hunt, J. C. R and Snyder, W. H., Experiments on stably and neutrally stratified flow over a model three-dimensional hill, *J. Fluid. Mech.*, **96** (1980) 671-714.

H18. Huxley, J., *Problems of relative growth*, Methuen (1932).

I1. Ishi, K., Analytical shape determination for membrane structures, *World Congress on Space Enclosures I*, Building Research Centre, Concordia University, Montreal (1976) 137-149.

I2. Iwamoto, K., *Go for beginners*, Penguin (1976).

J1. James, W. R., Fortran-IV program using double Fourier series for surface fitting of irregularly-spaced data, *Kansas Geological Survey Computer Constribution*, **5** (1966).

J2. Jeans, J. H., *Astronomy and cosmogony*, Cambridge University Press, Dover (1961).

J3. Jones, O., *The grammar of ornament*, Quaritch, London (1857, 1910).

J4. Junkins, J. L., Miller, G. W. and Jancaitis, J. R., A weighting function approach to modelling of irregular surfaces, *J. Geophys. Research*, **78** (no. 11) (1973) 1794-1803.

K1. Kasner, E. and Newman, J., *Mathematics and the imagination*, G. Bell, London (2nd edn, 1965).

K2. Kepes, G. (ed.), *Structure in art and science*, Studio Vista (1965).

K3. Kepes, G. (ed.), *Module, symmetry, proportion*, Studio Vista (1966).

K4. Klarner, D. A. (ed.), *The mathematical Gardner*, Wadsworth International (1981).

K5. Klein, F., Vergleichende Betrachtungen über neuere geometrische Forschungen, *Math. Annalen*, 43 (1903).

K6. Kober, H., *Dictionary of conformal representations*, Dover (1952).

K7. Kraus, H., *Thin elastic shells*, Wiley (1967).

K8. Krige, D., Two dimensional weighted moving average trend surfaces for ore evaluation, *Symp. on Mathematical Statistics and Computer Applications in Ore Valuation*, Johannesburg (1966) 13-38.

K9. Krumbein, W., Trend surface analysis of contour-type maps with irregular control-point spacing, *J. Geophys. Research*, **64** (1959) 823-834.

L1. Landau, L. D. and Lifschitz, E. M., *Fluid dynamics*, Pergamon (1959).

L2. Landau, L. D. and Lifschitz, E. M., *Theory of elasticity*, Pergamon (1959).

L3. Landau, L. D. and Lifschitz, E. M., *The classical theory of fields*, Pergamon (1962).

L4. Leopold, L. B. and Langbein, W. B., The concept of entropy in landscape evolution, *U.S. Geological Survey professional paper* 500A, (1962).

L5. Leopold, L. B. and Langbein, W. B., River meanders, *Scientific American*, Jan. (1966) 60-70.

L6. Levine, M. D., Feature extraction: a survey, *Proc. IEEE*, **57** (no. 8) (1969) 1391-1407.

L7. Lietzmann, W., *Visual topology*, Chatto and Windus (1955).

L8. Locher, J. and Escher, M. C., *The worlds of M. C. Escher*, Abrams, New York (1974).

L9. Loeb, A. L., The architecture of crystals (ref. K3) (1966) 38-63.

L10. Lord, E. A., The interaction of the radiation field with material objects, *Building and Environment*, **15** (1980) 167-179.

L11. Lynes, J. A., *Principles of natural lighting*, Elsevier (1968).

M1. Mackay, A. L., Generalised crystallography, *Izvještaj Jugoslavenskog centra za kristalografiju*, **10** (1975) 15-36.

M2. Mackay, A. L., Crystal symmetry, *Physics Bulletin*, No. (1976) 495-497.

M3. March, L. and Steadman, P., *The geometry of environment*, RIBA (1971).

M4. March, L. (ed.), *The architecture of form*, Cambridge University Press (1976).

M5. March, L., A Boolean description of built form (ref. M4) (1976) 41-73.

M6. March, L. and Earl, C. F. On counting architectural plans, *Environment and Planning b*, **4** (1977) 57-80.

M7. Marks, R. W., *The dymaxion world of Buckminster Fuller*, Reinhold (1960).

M8. Meltzer, B., Searle, N. H. and Brown, R., Numerical specification of biological form, *Nature*, **216** (1967) 32-36.

M9. Milne-Thomson, L. M., *Theoretical hydrodynamics*, Macmillan (5th edn, 1968).

M10. Mitchell, A. R. and Wait, R., *The finite element method in partial differential equations*, McGraw-Hill Kogakusha (2nd edn, 1972).

M11. Morley, F. and Morley, F. V., *Inversive geometry*, G. Bell, London (1933).

M12. Moore, E. H., On certain crinkly curves, *Trans. Amer. Math. Soc.*, **1** (1900) 72-90.

M13. Morse, P. M. and Feshbach, H., *Methods of theoretical physics*, McGraw-Hill (1953).

N1. Neal, B. G., *The plastic method in structural analysis*, Chapman and Hall (1956).

O1. Ore, O., *Theory of graphs*, American Mathematical Society (1962).

O2. Ore, O., *Graphs and their uses*, Random House (1963).

O3. Owen, W. S. and Niedermair, J. C., Geometry of the ship, *Principles of naval architecture* (J. Comstock, ed.), Society of Naval Architects and Marine Engineers (1967).

P1. Palmer, J. A. B., Computer science aspects of the mapping problem (ref. D3) (1975) 115-172.

P2. Peano, G., Sur un courbe, qui remplit toute un aire plane, *Math. Annalen*, **36** (1890) 157-160.

P3. Pearce, P., A minimum inventory maximum diversity building system, *2nd Int. Conf. Space Structures*, University of Surrey, (1975) 663-674.

P4. Pearce, P., *Structure in nature is a strategy for design*, M.I.T. Press (1978).

P5. Pfalz, J. L. and Rosenfeld, A., Computer representation of planar regions by their skeletons, *Comm. Assoc. Computing Machinery*, **10** (no. 2) (1967) 119-122.

P6. Pfalz, J. L., Representation of geographic surfaces within a computer (ref. D3) (1975) 210-230.

P7. Phillips, E. G., *Functions of a complex variable*, Oliver and Boyd (1961).

P8. Phillips, F. C., *An introduction to crystallography*, Longmans (4th edn, 1971).

P9. Penrose, R. The role of aesthetics in pure and applied mathematical research, *Bull. Inst. Math. Appl.*, **10** (no. 7/8) (1974) 266-271.

R1. Redish, K. A., *An introduction to computation methods*, The English University Press (1961).

R2. Richards, F. J., The geometry of phyllotaxis and its origin, *Symp. Soc. Exp. Biol.*, **2** (1948) 217-245.

R3. Richards, F. J., Phyllotaxis: its quantitative expression and relation to growth in the apex, *Phil. Trans. Roy. Soc. B.*, **235** (1951) 509-564.

R4. Robinson, J. E., Frequency analysis, sampling, and errors in spatial data (ref. D3) (1975) 78-95.

R5. Rosenfeld, A. and Kak, A. C., *Digital picture processing*, Academic Press (1976).

R6. Rushing, T. B., *Topological embedding*, Academic Press (1973).

S1. Schattenschneider, D., In praise of amateurs (ref. K4) (1981) 140-166.

S2. Schiff, L. I., *Quantum mechanics*, McGraw-Hill Kogakusha (2nd edn, 1955).

S3. Schrandt, R. G. and Ulam, S. (ref. B13) (1970) 234.

S4. Schwenk, T., *Sensitive Chaos*, Rudolf Steiner Press (1965).

S5. Seifert, H. and Threlfall, W., *A Textbook of Topology*, Academic Press (1980).

S6. Semple, J. G. and Kneebone, G. T., *Algebraic projective geometry*, Oxford University Press (1952).

S7. Semple, J. G. and Kneebone, G. T., *Algebraic curves*, Oxford University

Press (1959).

S8.   Siegel, K., *Structure and form in modern architecture*, Reinhold, New York (1962).

S9.   Smith, D. G., *Numerical solution of partial differential equations, finite difference methods*, Oxford University Press (2nd edn, 1978).

S10. Snow, R., Problems of phyllotaxis and leaf determination, *Endeavor*, Oct. (1955) 190-199.

S11. Southwell, R. V., *Relaxation methods in theoretical physics*, Oxford University Press (1946).

S12. Spinrad, R. J., Machine recognition of hand printing, *Information and Control*, 8 (1965) 124-142.

S13. Stacey, F. D., *Physics of the earth*, Wiley (2nd edn, 1977).

S14. Steadman, P., Graph-theoretic representation of architectural arrangement (ref. M4) (1976) 94-115.

S15. Steers, J. A., *An introduction to the study of map projections*, University of London Press (7th edn, 1949).

S16. Stevens, P. S., *Patterns in nature*, Penguin (1976).

S17. Stewart, I., The seven elementary catastrophes, *New Scientist*, 20 Nov. (1975) 447-454.

S18. Stiny, G., *Pictorial and formal aspects of shape and shape grammars*, Birkhäuser (1975).

S19. Stiny, G., Kindergarten grammars: designing with Froebel's building gifts, *Environment and Planning b*, 7 (no. 4) (1980) 409-462.

S20. Struik, D., *Differential geometry*, Addison Wesley (2nd edn, 1961).

S21. Sullenger, D. B. and Kennard, C. H. L., Boron crystals, *Scientific American*, July (1966) 96-107.

T1.   Thom, A. and Apelt, C. J., *Field computations in engineering and physics*, von Nostrand (1961).

T2.   Thom, R., *Structural stability and morphogenesis*, Benjamin (1975).

T3.   Thompson, D. W., *On growth and form*, Cambridge University Press (1917, abridged edn, 1961).

T4.   Töpfer, F., *Kartographische Generalisierung*, Hermann Hark, Leipzig (1977).

U1.   Uhr, L. (ed.), *Pattern recognition*, Wiley, New York (1966).

U2.   Ulam, S., Patterns of growth of figures: mathematical aspects (ref. K3) (1966).

V1.   Vajda, S., *Patterns and configurations in finite spaces*, Charles Griffin, London (1967).

V2.   Van Iterson, G., *Studien über Blattstellungen*, G. Fischer, Jena (1907).

V3.   Veblen, O. and Bussey, W. H., Finite projective geometries, *Trans. Amer. Math. Soc.*, 7 (1906) 241.

V4.   Vero, R., *Understanding perspective*, Van Nostrand Reinhold (1980).

V5.   Voderberg, H., Zur Zerlegung der Ebene in kongruente Bereiche in Form einer Spirale, *J. Ber. Deutch. Math. Verein*, **46** (1936) 229-231.

W1. Waddington, C. H., The modular principle and biological form (ref. K3) (1966) 20-37.

W2. Wagner, K., *Kartographische Netzentwürfe*, Bibliographisches Institut, Leipzig (1949).

W3. Walkers, N. V. and Bromham, J., *Principles of perspective*, Architectural Press (1970).

W4. Wardlaw, C. W., *Phylogeny and morphogenesis*, Macmillan (1952).

W5. Wardlaw, C. W., *Morphogenesis in plants*, Methuen (1968).

W6. Wathen-Dunn, W., (ed.), Models for the perception of speech and visual form, M.I.T. Press (1967).

W7. Wenninger, M., *Polyhedron models*, Cambridge University Press (1971).

W8. Weyl, H., *Symmetry*, Princeton University Press (1921).

W9. Whitten, E. H. T., The practical use of trend-surface analyses in the geological sciences (ref. D3) (1975) 282-297.

W10. Woodcock, A. E. R. and Poston, T., *A geometrical study of elementary catastrophes*, Springer (1970).

W11. Woolfson, M. M., *An introduction to X-ray crystallography*, Cambridge University Press (1970).

Y1. Yoeli, P., Compilation of data for computer-assisted relief cartography (ref. D3) (1975) 352-367.

Z1. Zeeman, E. C., *Catastrophe theory, selected papers*, Addison Wesley (1977).

Index